£21.00

23/31 Charlotte Street

London W1

6373961

THERMAL ENERGY CONSERVATION
Building and Services Design

THERMAL ENERGY CONSERVATION
Building and Services Design

by

J. W. WELLER
B.Sc. (Hons), Ph.D., C.Eng., M.C.I.B.S., M.C.I.O.B., M.Inst.E.

Adams Green & Partners, Consulting Engineers, Link House, Elmfield, Millbrook Road East, Southampton SO1 0HY, UK

and

A. YOULE
M.A., Ph.D. (Cantab).

School of Architecture, Portsmouth Polytechnic, King Henry I Street, Portsmouth PO1 2DY, UK

APPLIED SCIENCE PUBLISHERS LTD
LONDON

APPLIED SCIENCE PUBLISHERS LTD
RIPPLE ROAD, BARKING, ESSEX, ENGLAND

British Library Cataloguing in Publication Data

Weller, J W
Thermal energy conservation.
1. Architecture and energy conservation
2. Buildings—Energy conservation
I. Title II. Youle, A
693.8′32 NA2542.3

ISBN 0-85334-938-X

WITH 33 TABLES AND 45 ILLUSTRATIONS

Printed in Great Britain by Galliard (Printers) Ltd, Great Yarmouth

Preface

This book has been based on the practical and academic experience of the authors obtained since the beginning of the 1970s when the 'energy crisis' made itself felt in all walks of life. The book was written because people from widely different backgrounds expressed an interest in energy conservation in buildings and services but were unable to obtain literature which discussed this important topic from all aspects.

The book has two main aims within the subject area of the title. Firstly, it covers those scientific and technical topics relating to thermal aspects of building design in such a manner as to give the reader understanding and appreciation without undue complexity and detail. Secondly, it aims to cover the more practical aspects of building, services and energy consumption, from the technical decisions and predictions which must be made during design to discussing the processes by which the design team makes these decisions.

It is acknowledged that in attempting to cover such a wide range of topics in a concise fashion, detailed consideration of subject matter has invariably to be put aside. However, references to further sources of information have been quoted where appropriate, and readers should consider the book as providing an overall view of the subject area and a framework into which existing or further knowledge can be placed.

Consideration has not been given to the selection of sources of energy. The choice ranges from which type of conventional fuel is most appropriate to whether 'alternative' energy sources (e.g. sun or wind) should be employed. There are many other sources of information which give technical detail on this matter. Essentially the economic considerations discussed in Chapter 7 will be of importance in the selection while other issues, such as reliability of supply, availability in the future, environmental pollution, storage and handling problems, and public image should be taken into account.

Units are expressed in the book in the format usually used by engineers rather than in the strict SI notation, e.g. m/s rather than ms^{-1} and °C rather than K. The symbol k is used for thermal conductivity rather than λ; the latter was adopted by the CIBS in 1980.

The authors would like to thank Professor G. H. Broadbent and other colleagues at the School of Architecture, Portsmouth Polytechnic, and at Adams Green & Partners, Consulting Engineers, Southampton, for helpful discussions and continuous support in the preparation of the manuscript.

Particular thanks are due to E. E. Stanley, Principal Lecturer in Design Science of the School of Architecture, Portsmouth, for invaluable support and advice especially on technical aspects, and M. J. Green, Senior Partner at Adams Green & Partners for helpful contributions in the refinement of the manuscript.

Thanks are also offered to members of the Building Research Establishment (BRE) and Building Services Research and Information Association (BSRIA) who contributed much on technical matters.

All references to published work and all sources of illustrations are acknowledged where appropriate in the text.

Thanks are also due to all the typists who contributed to the preparation of the manuscript and to P. D. Jones for assistance in preparing the illustrations.

J. W. WELLER and A. YOULE

Contents

Chapter 1

Basic Aspects of Heat

1.1 HEAT AND ENERGY

Heat is a form of energy. This is well known, for energy in the form of fuel, e.g. coal and gas, is used to heat the buildings we occupy. The energy stored in fuels is chemical energy; this is released as heat energy when the fuel burns, to be transferred around a building or used, for instance, for cooking. This 'heat' energy is sensed by human beings, who endeavour to create conditions where the appropriate amount of heat energy is transferred between themselves and the surroundings so that they 'feel' thermally comfortable.

The basic unit of quantity of heat is the joule (J). This is a very small unit, for a domestic electric fire gives off typically 1000 J every second (10^3 J/s), i.e. one kilojoule per second (1 kJ/s) and the energy consumption of a building in a year is measured in units of 10^9 joules (i.e. one gigajoule, GJ). Usually the rate at which heat is produced, or lost, is of particular interest so that the joule per second, namely the watt (W), or more practically the kilowatt, kW (10^3 W), is used for describing heat gains or losses from buildings or outputs from heating units. A 1 kW fire running for 1 h (3600 s) gives out a total of 1000×3600 J, i.e. $3{\cdot}6 \times 10^6$ J. This quantity of heat is termed for convenience a kilowatt hour (kWh). Approximately 278 kWh is equivalent to 1 GJ (10^9 J). The kWh or GJ can be used to cost the consumption of energy.

The energy requirement of a building is inherently linked to the thermal performance. Thus decisions during building design which either influence or are directly related to obtaining appropriate thermal conditions in a building must have a bearing on the amount of energy consumed by the building. If this consumption is to be reasonably controlled an understanding of the thermal aspects is a clear prerequisite.

1.2 FUNDAMENTAL DESCRIPTION OF HEAT

Heat energy is an energy of movement, namely the movement of atoms and molecules either freely and randomly in a space, as with a gas, or vibrating about a fixed point as in a solid. This is well illustrated by simple mechanical action being translated into heat, for instance when the action of filing a piece of metal causes the atoms to become more agitated and the metal feels warmer.

Conversely, when a gas is heated the molecules begin to move about more rapidly. So if a tin with a lid is heated, the lid will eventually burst off because the air molecules inside are hitting it more frequently and with more speed. The net result is an increased force on the inside of the lid (i.e. the pressure in the tin increases).

Most materials change their state of matter from solid to liquid to gas as their heat content rises. The state of matter that exists for a material under normal 'room' conditions is a balance between the thermal movement forces and the inherent binding forces between the atoms or molecules of the material. If the binding forces are dominant then the material is a solid while if the binding forces are weak the material will be a liquid or gas as the thermal forces will dominate. At very high heat contents the binding forces of all materials are overcome and gases prevail, while at very low heat contents all materials turn to solids. In the extreme case of no heat at all, all movement due to heat is removed, there is no atomic vibration in the material and the condition is termed absolute zero. This is the theoretical base level of all temperature measurement.

1.2.1 Scales of Temperature

Temperature is an artificial concept derived to express in simple terms the amount of heat energy that exists in a material. The zero heat energy point would represent zero temperature on some scale, this scale being known as the Absolute or Kelvin scale (units K), and the units of the scale would represent defined amounts of increasing heat content. Other scales take commonly occurring 'heat contents' as bases, e.g. the heat content of ice mixed with water in equilibrium as a base of zero, and arbitrarily 100 units between this base and another fixed point, e.g. the heat content of water and steam in equilibrium. This is the Celsius scale which is commonly called the Centigrade scale (°C), while other fixed points generate the Fahrenheit scale (°F). Zero on the Celsius scale is equivalent to 273 K.

Values of temperature are based on some easily measured property of a material which changes with heat content. Most common in everyday use is the increase in length of a material, e.g. a column of mercury or alcohol in a glass tube. The change in length is graduated as a scale of temperature. Many other properties change with heat content and can be similarly used. Examples are electrical conductivity, voltages generated when dissimilar metals are connected (the thermocouple effect) and light given off from hot objects.

1.2.2 The Transfer of Heat

There is a basic property of Nature (expressed in the Laws of Thermodynamics) which states that if there is a region of high heat content, i.e. high temperature, and one of low heat content, i.e. low temperature, then there will always be a spontaneous tendency or a driving force for heat to travel from the higher to the lower region. In other words heat is always trying to even itself out so that if allowed, everything would eventually arrive at the same temperature. This effect is similar to water finding its own height and flowing down slopes until it settles at a uniform level. Thus thermal energy is always on the move, tending to dissipate itself.

How does this heat transfer occur? There are three transfer mechanisms, namely radiation, convection and conduction, and a fourth, rather indirect method, based on the change of state of materials. Each has its own particular characteristics and must be understood before the thermal performance of buildings can be fully appreciated.

1.3 THERMAL RADIATION

Radiation is perhaps the most fundamental and important heat flow mechanism as it is in this way that the planet Earth receives most of its energy from the sun. The energy that mankind has inherited through fossil fuels is simply stored energy derived originally from the sun's radiation.

> *Radiation is the flow of energy by the passage of electromagnetic waves through a vacuum or transparent material.*

The term *electromagnetic waves* must first be clarified. The physical understanding of it is a complex process; the term is best appreciated by looking at examples. The nature and description of the electromagnetic radiation is dependent on the wavelength involved, as shown in Fig. 1.1. Hence visible light, X-rays and radio waves are all fundamentally the same, but their manifestation occurs in different ways. Each type of radiation transfers energy, the nature of which is demonstrated when the radiation interacts with materials. For example, X-rays can disrupt the structure of molecules such as in human tissues; light rays can cause electrons to be released from certain materials generating an electric current, e.g. in a

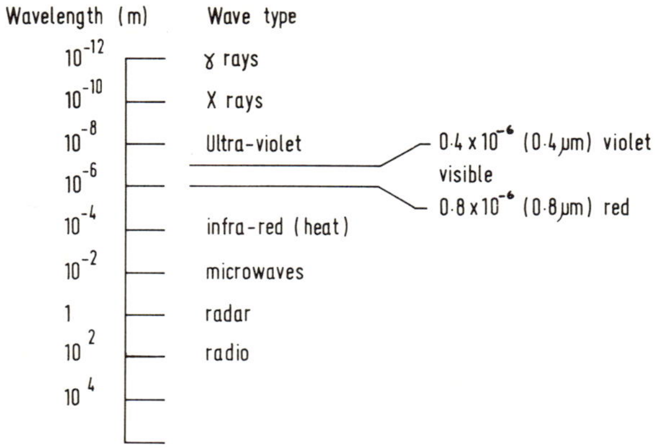

Fig. 1.1. The electromagnetic spectrum showing radiation type and corresponding wavelength range.

photocell; radio waves can generate electron flow in metals as used in aerials for collecting radio signals.

One particular section of the electromagnetic spectrum when absorbed by materials converts its energy directly to the thermal form, i.e. the radiation causes atoms and molecules to vibrate. This section is known as the *thermal* or *infra-red* band of the spectrum, as identified in Fig. 1.1, and occurs immediately adjacent to the red part of the visible spectrum—hence the term *infra-red.* For this reason *thermal radiation* can be considered in many ways to behave in a similar fashion to light radiation. Hence, just as light waves emanate from an electric bulb, so do *infra-red* rays. Infra-red wavelengths range from 0·8 micrometres (μm) to approximately 100 μm (1 μm = 1 millionth of a metre).

Many aspects of the behaviour of electromagnetic radiation, in particular the energy it carries and the materials it will travel through, depend on the wavelength of the radiation. Thus infra-red radiation has its own particular characteristics based on the wavelengths involved, and these characteristics must be examined in detail before the properties and implications for building design can be exploited.

1.3.1 Quantity of Thermal Radiation Emitted from an Object

All objects are emitting thermal radiation. This fact is often not fully appreciated; one normally only senses and detects 'heat' radiation given off from very hot objects. Thus a coal fire 'feels' hot when standing in front of it and the sun warms people on the top of ice covered mountains. This is because these hot objects are giving off large amounts of thermal radiation which can be sensed directly and instantly, but in fact all objects experienced in everyday life are emitting such radiation, but at lower intensities.

The amount of radiation given off from an object is clearly dependent on the temperature of the object. It is also dependent on the size of the object. Three bars of an electric fire give off more heat than one bar. Finally, it depends on a further factor, known as the emissivity of the surface, which in simple terms means that, for instance, a black object will give off more radiation than a similar polished metallic object at the same temperature. This term will be explained further in Section 1.3.3.

These factors are combined to quantify the rate of heat transfer by radiation (Q_r) from an object.

Q_r, measured in watts (W), is dependent on:

(a) the absolute temperature, T, (measured in Kelvin) of the object, raised to the fourth power. The absolute temperature must be used as the relationship holds only for a scale based on positive values, i.e. starting at absolute zero.
(b) the surface area, A, of the object (in m^2).
(c) the emissivity, ε, of the surface (a dimensionless parameter).

Hence:

$$Q_r = \sigma \varepsilon A T^4$$

where σ is the constant of proportionality, known as the Stefan–Boltzmann constant, with a value of $5{\cdot}67 \times 10^{-8}\ \mathrm{W m^{-2} K^{-4}}$ (the units of this constant result simply from the units and powers of the other parameters in the equation).

To illustrate the magnitudes involved, this equation is used to calculate the quantity of radiation emitted for unit surface area (i.e. $A = 1\ \mathrm{m}^2$) from an object at 300 K (i.e. 27 °C) and at four times this absolute temperature, i.e. 1200 K (927 °C). For simplicity the emissivity, ε, is assumed to be 1. The values obtained are respectively 460 W and 117 600 W, the latter being 256 times (i.e. 4^4) as great because of the T^4 relationship.

The value of 460 W radiated from 1 m^2 of surface at 27 °C may appear to be high, but it must be appreciated that if all objects in a room are at this temperature then they will not only be giving off this amount of radiation but also receiving it, i.e. there is no net radiation gain to any object (see Section 1.3.4).

High temperature sources such as coal fires or electric bar radiators will be emitting radiation of similar intensity to the 1200 K source determined. However, the total area involved is usually very much less than 1 m^2.

1.3.2 Wavelength of Radiation

The wavelength of electromagnetic radiation has significant implications for the properties of the radiation. This has been illustrated by the differences in properties between short wavelength radiation, e.g. X-rays, and long wavelength radiation, e.g. radio waves. Even small shifts in wavelength can produce important changes in properties. With visible radiation the obvious effect is the change in perceived colour of the radiation, while other property changes also occur, for instance in the refraction of light when it passes through glass.

The thermal radiation given off from an object is not of a specific wavelength, but covers a continuous range of values where the energy of the radiation varies with the wavelength. The energies and wavelengths involved depend on the temperature of the object; with increasing temperature there is an increase in energy (as has already been established in Section 1.3.1) but there is also a shift in the overall wavelengths to shorter values.

These effects are summarised by plotting intensity against wavelength for different temperature values, as shown in Fig. 1.2. The curves for higher temperature objects have a greater area underneath them, indicating increased intensity, and the peak of the curve shifts to the left, i.e. to shorter wavelength values.

A simple manifestation of this is the behaviour of an iron poker as it heats up in a fire. Firstly it glows dull red; as it gets hotter the appearance turns to a brighter red, to orange and to white, demonstrating that the visible radiation given off from the iron is shifting to smaller wavelengths. The visible radiation is, in fact, the extreme left hand edge of the curves in Fig. 1.2. The bulk of the radiation is not visible as it is in the infra-red region. These effects can be confirmed experimentally by the collection and measurement of the radiation.

The wavelength characteristics of infra-red radiation may at first appear to be abstract in nature, but they will be seen to have important implications for properties such as the transmission of radiation through glass, and the reflection and absorption properties of surfaces.

1.3.3 Emissivity and Absorptivity

The term emissivity (ε) has already been introduced (Section 1.3.1) in relation to the quantity of radiation emitted from a surface. An object at a

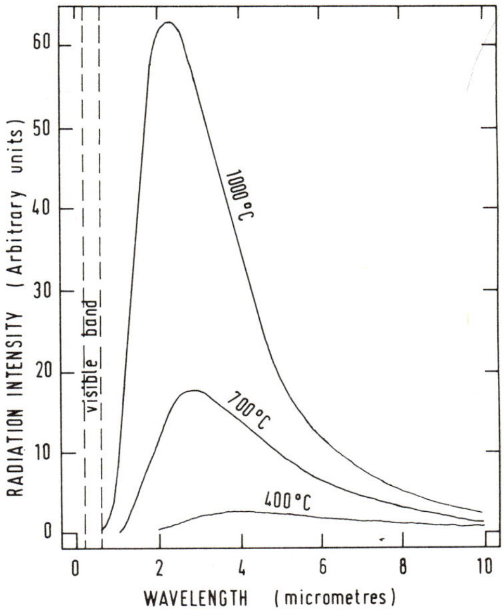

Fig. 1.2. Graph of emitted radiation intensity versus wavelength for objects at different temperatures.

given temperature can emit a theoretical maximum amount of radiation, and if it did so its emissivity would have the value 1, and the object would be known as a perfect black body. However, in practice an object emits some fraction of the theoretical maximum, and this fraction is the emissivity value.

Emissivity (ε) *is defined as the ratio of the thermal radiation emitted from unit area of a surface to the radiation from unit area of a full emitter ('black body') at the same temperature.*

Emissivity values are essentially related to the colour and nature of surfaces. A typical polished metallic surface has a value of about 0·1. Thus a polished kettle would emit about one-tenth of the radiation emitted by a matt black one, when both are full of boiling water; i.e. a polished kettle is thermally more efficient than a dull one as less radiant heat is lost from its outer surface.

An object not only emits radiation but absorbs it. The quantity it absorbs compared to the total radiation falling on it is known as the absorptivity of the surface, and again the value depends on the surface colour and nature.

Absorptivity (α) *is defined as the ratio of radiant energy absorbed by an object to the total black body radiation falling on it.*

That energy which is not absorbed is reflected from the surface. The reflectance value (r) *is hence equal to* $(1 - \alpha)$.

For uniform temperature conditions absorptivity and emissivity values are equal. Thus an object at 30 °C placed in surroundings of temperature 30 °C will absorb and emit equal quantities of radiation. If this was not the case the object or the surroundings would increase in temperature and the fundamental heat transfer law (the evening out of temperature) would be disobeyed.

However, many surfaces have absorptivity and emissivity values which are dependent on the radiation characteristics, i.e. on the temperature of the surface producing the radiation. For example, the absorptivity value of a white surface for incident radiation from objects at 30 °C is about 0·9, but for radiation from the sun (i.e. of surface temperature of about 10 000 °C) the value is about 0·4. Surfaces which do not have this temperature dependence are known as 'grey' bodies.

Examples of values of emissivity and absorptivity for low temperature radiation compared to absorptivity for solar radiation are given in Table 1.1. (Emissivity values for solar radiation temperatures are not relevant in the context of the fabric of buildings as materials do not reach such temperatures.)

These values have important implications in building design. For instance, in countries which have intense solar radiation conditions,

Table 1.1 Emissivity and absorptivity values for some common surfaces

Surface	Emissivity (ε) or absorptivity (α) for ambient conditions ($\sim$20 °C)	Absorptivity (α) for solar radiation (radiation source at $\sim 10^4$ °C)
Matt black	0·95–1·00	0·85–1·00
Dark brick	0·85–0·95	0·65–0·80
White stone	0·85–0·90	0·40–0·50
Dull metal	0·20–0·30	0·40–0·65
Polished metal	0·02–0·15	0·10–0·30

absorption }
emission } 0.85-0.90

painting the outside of buildings white ($\alpha = 0{\cdot}4$ to $0{\cdot}5$) has an important effect in terms of lowering the solar radiation absorbed by the building fabric during the day and hence reducing the building heat gains. (Ideally the external fabric would have a polished metallic surface ($\alpha = 0{\cdot}1$–$0{\cdot}3$) but this is often an impractical building surface finish.) However, the white surface has a relatively high emissivity value, 0·85–0·9 for ambient temperature conditions, and hence this can help cool the building at night. On the other hand, for hot water radiators in a house there is little difference in emissivity values of the radiators if they are painted matt black or any other colour (emissivity values ranging from 1 to 0·75) which is fortunate from the aspect of appearance. However, a hot water radiator finished with a polished metallic surface would have a poor thermal radiation performance. With a kettle, though, as has been previously stated, this is a purpose designed finish to minimise radiant heat loss.

1.3.4 Radiation Exchange

It has been established that all objects emit thermal radiation, the quantity depending on their temperature. It must follow therefore that all objects must be receiving radiation from their surroundings. In equilibrium an object will be emitting as much radiation as it is receiving and if this is the only transfer process occurring its temperature will remain constant. However, if a warm object is surrounded by a set of cold walls, then there will be an overall loss of radiation from the object as it is emitting more than it is receiving, and its temperature will drop.

Thus when a person (a warm object) is standing in a room with cold walls (e.g. expanses of glass on a cold day) there is a net radiation loss from the person to the walls, and thermal comfort might not be maintained. (There would also be other loss processes occurring which have yet to be discussed.)

Similarly, if a person is standing in an office surrounded by tinted glass which has been heated up by the absorption of solar radiation to temperatures in excess of the body surface temperature, then there will be a net gain in radiation to the person, who might become uncomfortably warm.

Before aspects of the exchange of radiation between an object and its surroundings can be considered further, the concept of *mean radiation temperature* must be established. For a very simple example of an object situated at the centre of a cubical room, the radiation that the object receives is dependent on the surface temperatures of the surrounding walls and their emissivities. If these are all typical building surface finishes of high emissivity (0·85–1·0) and they are all at the same temperature, then the radiant temperature seen by the object will be this temperature value. However, if the surfaces are of varying temperatures then the average radiant temperature seen by the object will be a function of the areas and the temperatures involved. This temperature is known as the mean radiant temperature, specified by t_{mrt}, and can be calculated to an accuracy sufficient for most purposes by:

$$t_{mrt} = \frac{a_1 t_1 + a_2 t_2 + \cdots}{a_1 + a_2 + \cdots}$$

where $a_1, a_2 \ldots$ are the areas involved and $t_1, t_2 \ldots$ are the corresponding surface temperatures. t_{mrt} may be quoted in usual temperature units (°C) rather than absolute units as it represents an average surface temperature. The symbol t (small t) is adopted here for temperatures other than absolute temperature values.

In practice, the geometry involved is complex, and other radiant sources such as light fittings and people are present, and the calculation of the mean radiant temperature becomes complex (see, for example, [1.1]). An estimate of the mean radiant temperature can also be obtained from measurements using a globe thermometer, i.e. a thermometer inside a black sphere, further described elsewhere (e.g. [1.2]).

Having established the concept of mean radiant temperature the interchange of radiation between an object and its surroundings can be examined. Consider an object at absolute temperature T_1, in surroundings of absolute mean radiant temperature T_2. The radiation, Q_{r1}, per unit area (i.e. $A = 1$) emitted from the object is

$$Q_{r1} = \sigma \varepsilon T_1^4 \qquad \text{(Section 1.3.1)}$$

The radiation, Q_{r2}, absorbed by the object per unit area is

$$Q_{r2} = \sigma \alpha T_2^4$$

If T_1 and T_2 do not differ greatly, i.e. for most conditions except direct solar radiation gain, then α and ε will have the same value (Section 1.3.3).

Hence the net radiation exchange of the object to its surroundings, Q_{r1-2}, is given by

$$Q_{r1-2} = \sigma \varepsilon (T_1^4 - T_2^4)$$

In other words the net exchange of radiation between objects is proportional to the difference between their absolute temperatures to the fourth power. This is known as Prévost's Theory of Exchanges.

Example 1.1

An object of $\varepsilon = 0{\cdot}9$ at 20 °C (293 K) is in surroundings of mean radiant temperature of 10 °C (283 K). What is the rate of heat loss by radiation per unit area from the object?

$$Q_{r1-2} = 5{\cdot}67 \times 10^{-8} \times 0{\cdot}9 \times (293^4 - 283^4)$$
$$= 48{\cdot}8 \text{ W}$$

i.e. the heat loss by radiation is approximately 50 W/m².

A simplified approximate alternative method of expressing radiation exchange introduces a term known as the radiation coefficient, h_r. The heat exchange for unit area is now given by

$$Q_{r1-2} = \varepsilon h_r (t_1 - t_2)$$

The fourth power component of the temperature term has been removed, and hence the temperatures may be quoted in degrees Centigrade and the symbol t is used. The radiation coefficient contains the Stefan–Boltzmann constant and its value depends on the temperatures involved, compensating for the simplified temperature component. For typical room conditions

$$h_r = 5{\cdot}7 \text{ W/m}^2\,°\text{C}$$

Taking Example 1.1 where $\varepsilon = 0{\cdot}9$ and the temperatures were 20 °C and 10 °C the expression gives

$$Q_{r1-2} = 0{\cdot}9 \times 5{\cdot}7 \times (20 - 10)$$
$$= 51 \text{ W/m}^2$$

This value is close to that previously obtained.

The radiation coefficient approach allows for simpler calculations. It also gives a heat transfer expression similar to that for convection transfer (Section 1.4.1) and is of value in discussing the transfer of heat at surfaces (Section 1.6.1).

1.3.5 Transmission of Thermal Radiation through Glass

Incident infra-red radiation onto a glass surface with air on either side will be partially reflected, partially absorbed and partially transmitted. The relative effects of each process will depend on a number of factors.

Reflection. This will depend on the angle of incidence of the incoming radiation to the surface. For angles between 0° and 40° the reflection value is about 8%. Above about 60° the quantity reflected increases rapidly to 100%. These effects are summarised in Fig. 1.3.

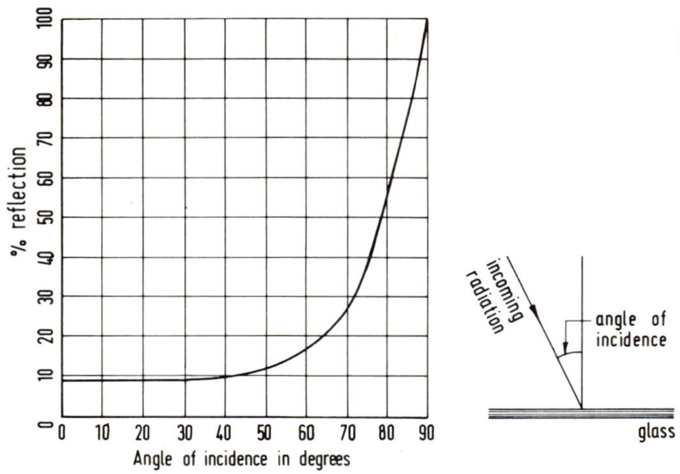

Fig. 1.3. Percentage reflection versus angle of incidence for radiation striking a glass surface (from Turner [1.3], reproduced by permission of Pilkington Flat Glass Ltd).

An example of these effects is the difficulty experienced in looking into shop windows from a side-on position; most of what one sees is reflections from down the street. It is also the reason why angled sections of glazing on a glass clad building can create pronounced visual effects through contrasting reflections.

Absorption. The quantity of infra-red radiation absorbed by glass depends on the radiation wavelength, the thickness, purity and cleanliness of the glass and on the angle of travel through the material. For typical thicknesses of clear glass for solar radiation striking the glass at zero angle of incidence, absorption of about 12 % of the total incoming radiation occurs. Glasses with a mild colouring, i.e. body tinted glasses absorb typically 50 %. Some 'solar control' glasses absorb up to 80 % of incoming infra-red radiation, but they also absorb a high percentage of visible radiation, so their performance as transparent materials begins to become questionable.

The absorbed radiation has the effect of heating up the glass; panes of clear and tinted glass similarly sited in sunlight can be felt by touch to have very different temperatures. This can cause significant changes to the mean radiant temperature in the space. The wavelength dependence of absorption is seen in the transmission characteristics of glass as follows.

Transmission. The combined effects of reflection and absorption are demonstrated by examining the resultant transmitted radiation. This is summarised for two types of glass in Fig. 1.4 by plotting the percentage of radiation transmitted against the wavelength of the radiation in the optical and infra-red regions.

It can be seen that little radiation is transmitted below 0·35 μm which is the ultra-violet region. This explains why a suntan (which depends on ultra-violet rays being absorbed by the skin) is not obtained when sunbathing

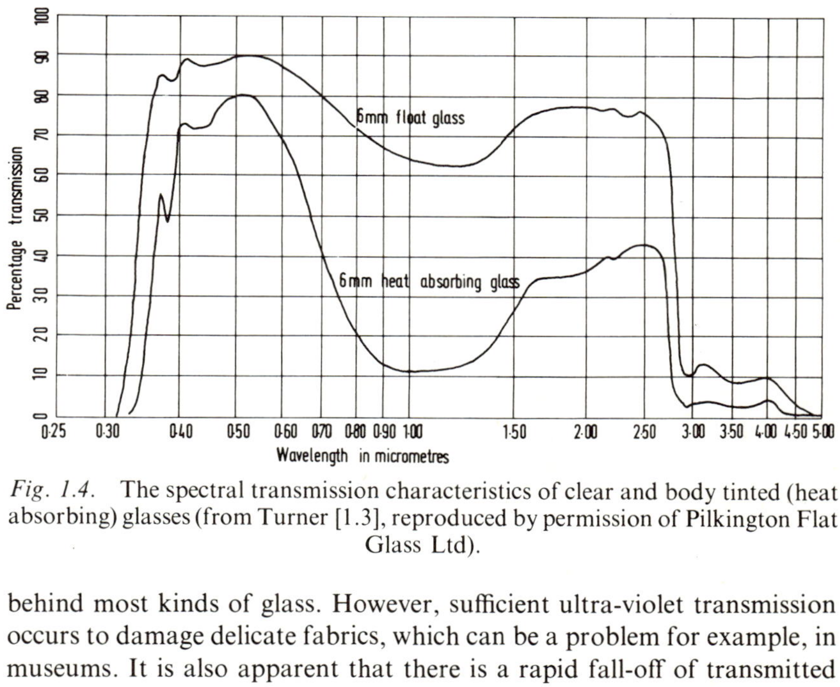

Fig. 1.4. The spectral transmission characteristics of clear and body tinted (heat absorbing) glasses (from Turner [1.3], reproduced by permission of Pilkington Flat Glass Ltd).

behind most kinds of glass. However, sufficient ultra-violet transmission occurs to damage delicate fabrics, which can be a problem for example, in museums. It is also apparent that there is a rapid fall-off of transmitted radiation at 2·5 to 3 μm. In other words glass allows little of the long wave infra-red radiation to pass through it. The significance of this becomes apparent in the following section.

1.3.6 The Greenhouse Effect

The overheating of highly glazed office buildings is an unfortunate phenomenon of our time. The principle applying, though, has been used to advantage historically in horticulture and home gardening (e.g. [1.4]). The effect of 'trapping' the sun's heat within a glazed area is readily explained now that the fundamental properties of heat radiation have been established.

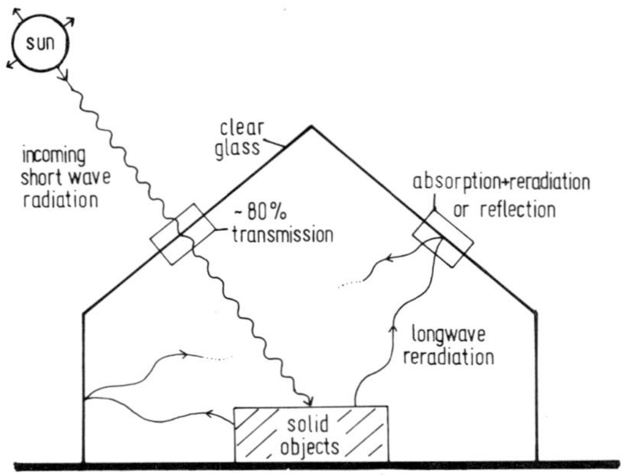

Fig. 1.5. The 'greenhouse' effect.

The sequence of events occurring, often known as the 'greenhouse' effect, is summarised in Fig. 1.5.

The effect is based on two principles already established.

(a) The wavelength characteristics of thermal radiation emitted from an object depend on the temperature of the object. The hotter the object the shorter the wavelength values. The short wave nature of solar radiation arriving at the ground (i.e. as affected by the absorption in the atmosphere) is given in Fig. 1.6.

(b) Glass allows high transmission of short wave radiation but little transmission of long wave radiation.

The sequence of events that occurs in the greenhouse effect is as follows:

(a) Incoming short wave intense radiation from the 'hot' sun strikes at low angles of incidence on the glass surfaces. The majority of this radiation (70–80%) is transmitted through the glass. The remainder is reflected at the surface or absorbed by the glass.

(b) The transmitted radiation falls onto objects within the space and is absorbed either initially or after reflection and scattering. The absorption has the effect of heating up objects in the space above the ambient air temperature.

(c) The heated objects are also emitting radiation, but as they are at relatively low temperatures the radiation is long wave in nature. Glass behaves as an opaque material to this long wave radiation, with an absorptivity value of approximately 0·9. Thus when this

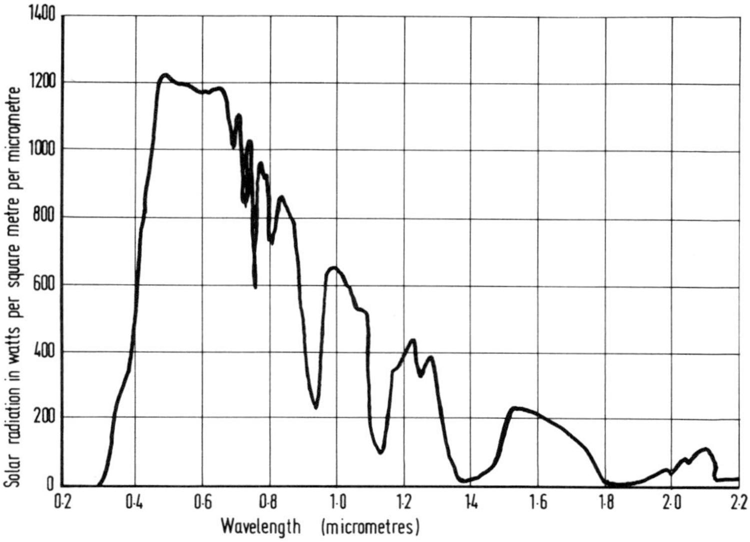

Fig. 1.6. Radiation intensity versus wavelength for solar radiation at the earth's surface. High absorption at certain wavelengths is due to specific components of the earth's atmosphere (from Turner [1.3], reproduced by permission of Pilkington Flat Glass Ltd).

radiation hits the inside of the glass it is mainly absorbed, causing the glass to warm up and reradiate back into the space and externally, while some of the radiation is reflected back into the space directly.

(d) The net effect is an overall heating up of the space, as the radiation has been changed in characteristic and is effectively 'trapped'.

The above explanation considers the main events occurring in radiation terms, but has not considered complexities arising due to conduction and convection effects. A fuller description can be found in [1.3].

The greenhouse effect has become widely recognised in the architectural world since some of the earlier problematic extensively glazed buildings. However, glass is a valuable material for external cladding and to maintain its application many types of 'solar control' glasses have been developed [1.3]. Rather optimistic hopes for solar control in design problems are sometimes attached to such glasses and the heating of the glass itself causing radiant heating into the space is frequently overlooked. Further aspects of solar control in relation to glazing are discussed in Sections 3.5 and 5.2.

1.3.7 Radiation Fall-Off with Distance from a Source

Radiation sources are widely used for the heating of occupants of buildings. There are also many unwanted sources of radiation which produce uncomfortable working conditions usually through overheating, but cold surfaces can also act as radiation 'sinks'. Such sources require control or correct application. The understanding of the control requires an appreciation of the way in which the intensity of radiation falls with increasing distance from a source. The general effect is very familiar; one stands back from a coal fire when the heating becomes excessive.

The behaviour is best understood by dividing sources into three simple categories, namely point, linear and planar. In practice one rarely has such simple sources, but a general appreciation of the differences is usually all that is required. In the explanations, sources at higher temperatures than their surroundings are considered.

(a) *Point source*. Radiation spreading uniformly in all directions from a point source such as a light bulb reduces in intensity considerably with distance from the source.

Consider the point source being at the centre of a sphere of 1 m radius. The radiation arriving at the surface of the sphere from its centre will be at a certain intensity. If this sphere is now replaced by a sphere of 2 m radius the same amount of radiation is now spread over the surface of this larger sphere, the area of which is four times the surface area of the smaller sphere (as the surface area of a sphere depends on the square of the radius). Hence the radiation intensity is only one quarter of that of the smaller sphere.

From this it can be seen that in general the radiation intensity per unit area is decreasing with the square of the distance from the source.

(b) *Line source* (*e.g. a heating pipe*). In this case there is now not one position (i.e. point) which is contributing to the radiation intensity at a given distance away from the source but a multitude of points along the line. The resulting effect can be shown to have the simple relationship, that the radiation intensity decreases directly with distance from the source, i.e. the intensity is inversely proportional to distance.

(c) *Planar source* (*e.g. a large kiln*). In this case it can be shown that there is no fall-off of radiation intensity with distance from the source, apart from a small reduction due to absorption by the atmosphere. This relationship holds so long as the source is large compared to the observer.

These effects are summarised in Fig. 1.7.

Examples of the applications of these effects arise in overheating problems in industry. If the heat source is small and localised (e.g. an arc flame) then an appropriate way of minimising the effect is to move away, as the intensity falls off rapidly with distance. With a cylindrical source this measure is not so effective and with a planar source (e.g. a large furnace, or a

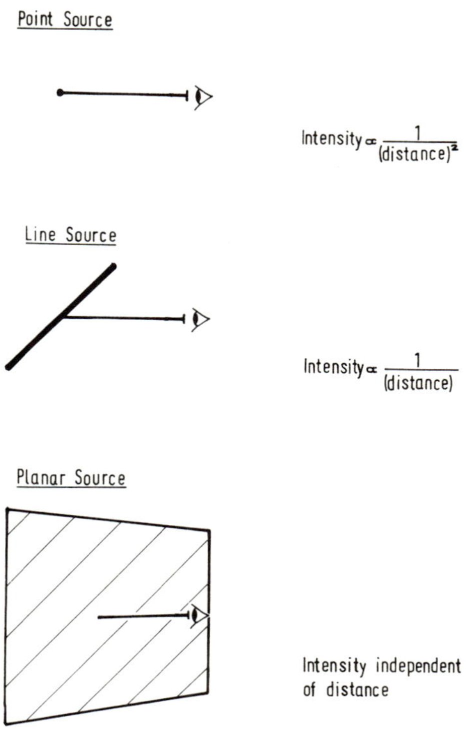

Fig. 1.7. Fall-off of radiation with distance from a point source, a line source and a planar source.

steel smelting plant) distance is of no value, and other measures must be taken, for instance the incorporation of radiation shields and reflectors (or the provision of fans for increased air movement).

When radiation sources are to be used for heating purposes then the main consideration is usually to ensure that sufficient intensity of radiation falls on the objects to be heated, so the reverse arguments to the above apply (see Section 5.3.3).

1.4 CONVECTION

When a fluid (e.g. air) is warmed it expands (through more activity being imparted to its molecules), its density decreases and hence it rises as it is buoyant relative to the surrounding fluid.

Convection is the transfer of heat by the bulk movement of a fluid (gas or liquid) created by thermal expansion effects or induced artificially.

A clear example of convection currents is the vigorous vertical flow of flames and smoke from an open fire. The intense heat generates rapid natural convection currents. However, the effect still applies but to a much lesser extent, to objects only a few degrees warmer than the air surrounding them. The opposite effect also occurs around a cold object, where cooled air tends to fall, for instance with air falling from a cold window surface.

1.4.1 Natural Convection

In the simplest terms the rate of heat lost through natural convection from an object hotter than the surrounding air is directly dependent on the temperature difference between the object and the air. Thus, if an object is at 100 °C and the air at 20 °C (temperature difference 80 °C) the heat loss will be eight times as great as for an object at 30 °C (temperature difference 10 °C).

However, the loss effect is also related to the size and shape of the object as the pattern of air movement created by convection is dependent on these factors.

Take, for example, a warm flat surface in varying orientations. If the surface is positioned horizontally with the warm side upwards then convection currents will rise immediately from the surface and heat transfer from the plate to the air will take place effectively, assuming replacement air can flow in from the sides. When the surface is vertical, air currents will flow readily upwards but the air becomes warmer as it moves up the surface, i.e. the temperature difference between the plate and the air will decrease and hence the heat loss will be reduced. Finally, with the surface horizontal and the warm side facing downwards, vertical air movement is impeded and the temperature difference decreases, reducing further the heat loss process. These effects are illustrated in Fig. 1.8. These arguments can be appropriately reversed for heat gains from surrounding air to colder surfaces. Thus in general terms the size, shape and orientation of an object will affect the air flow pattern created by natural convection, and hence the rate of cooling or heating of the object.

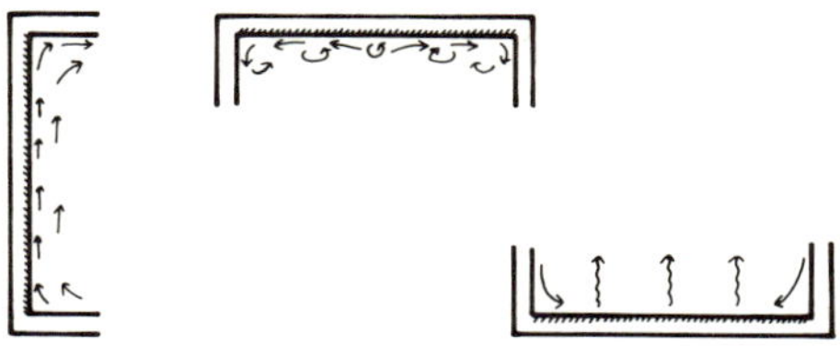

Fig. 1.8. Representation of convection currents from flat surfaces in three orientations, where the surface is warmer than the surrounding air.

The analysis of heat transfer by convection is complex, but a simplified expression for the process is given by:

$$Q_{cv} = h_c \times A \times (t - t_a) \text{ or } \Delta t$$

$$\underset{(\mathrm{W})}{\text{Heat transfer by convection}} = \underset{(\mathrm{W/m^2\,^\circ C})}{\text{Convection coefficient}} \times \underset{(\mathrm{m^2})}{\text{Area of surface}} \times \underset{(^\circ\mathrm{C})}{\left\{\text{Temperature of object} - \text{Temperature of air}\right\}}$$

Thus, given a variety of surfaces where Δt values are all identical, the heat transfer by convection from each surface will vary according to the h_c value. This value is dependent on the shape and orientation of the surface, which as already described affects the direction and speed of the air flow over the surface and localised temperature differences. h_c is also dependent on the overall Δt value, as this will affect the air flow pattern in terms of laminar or turbulent characteristics of the flow.

For typical conditions in internal rooms of moderate heights h_c has the following values:

Horizontally up surfaces	$h_c = 4{\cdot}3\ \mathrm{W/m^2\ ^\circ C}$
Vertical surfaces	$h_c = 3{\cdot}0\ \mathrm{W/m^2\ ^\circ C}$
Horizontally down surfaces	$h_c = 1{\cdot}5\ \mathrm{W/m^2\ ^\circ C}$

These values match the qualitative effects already described and illustrated in Fig. 1.8.

Example 1.2

What is the heat lost per unit area by convection from the face of a vertical wall of temperature 20 °C when the adjacent air temperature is 10 °C?

$h_c = 3{\cdot}0\ \mathrm{W/m^2\ °C}$ for a vertical surface; $\Delta t = 10\,°\mathrm{C}$; $A = 1\ \mathrm{m^2}$.

$$Q_{cv} = h_c A\, \Delta t$$

Substituting the values:

$$\begin{aligned} Q_{cv} &= 3{\cdot}0 \times 1 \times 10 \\ &= 30\ \mathrm{W} \qquad \text{for } 1\ \mathrm{m^2} \end{aligned}$$

i.e. the heat loss from the surface is approximately 30 $\mathrm{W/m^2}$.

It is of interest to note that for an object under the same temperature conditions the net rate of heat loss by radiation is approximately 50 $\mathrm{W/m^2}$ (as determined in Example 1.1, Section 1.3.4), which is nearly twice the rate for natural convection.

What is the effect if the temperature of the surface is raised to 60 °C, i.e. the surface temperature that may occur for a hot water 'radiator'?

Δt is now 50 °C.

$$Q_{cv} = 3 \times 50 = 150\ \mathrm{W/m^2}$$

i.e. such a surface gives off approximately 150 $\mathrm{W/m^2}$ by convection into the adjacent space.

However, a more thorough analysis would show that the convection coefficient increases with the temperature difference involved and the heat loss would be greater than that determined (see e.g. [1.5]).

1.4.2 Forced Convection

If air or liquid is *forced* over a hot object, rather than being allowed to flow naturally, then the heat lost from the object is increased. This is commonly experienced by the cooling effects of winds and the desire to 'get out of' the wind to keep warm, even though the air temperature of the wind and the still air are the same.

In the simplest terms the extra cooling effect depends on the velocity, v, of the moving air, raised to the power of $\frac{1}{2}$ (i.e. the square root of v). Thus air flowing at 16 m/s ($v^{1/2} = 4$) will have twice the cooling effect of air flowing at 4 m/s ($v^{1/2} = 2$). The effects of forced convection are very important in discussing thermal comfort (Section 2.2.3) and in considering air movement in buildings, and resultant heating and cooling effects (Chapter 4).

The velocity component in forced convection is incorporated into the convection heat loss equation through the convection coefficient h_c. The original equation applies, i.e.

$$Q_{cv} = h_c A \, \Delta t$$

but now h_c is velocity dependent and can be considered to have the values given by the following expressions:

$$h_c = 5{\cdot}8 + 4{\cdot}1v \qquad v < 5\,\mathrm{m/s}$$
$$h_c = 7{\cdot}8v^{0{\cdot}8} \qquad v > 5\,\mathrm{m/s}$$

The above values of h_c are clearly greater than those for natural convection while the temperature difference now has little effect on the h_c value because the forcing of the air influences the transfer rather than the temperature density effects. The forced h_c value is not quoted for different surface orientations as the air is being artificially directed at the surface rather than rising vertically in a natural manner.

Insertion of typical values of v in these equations will demonstrate the general velocity dependence which approximates, particularly for low values of v, to the $v^{1/2}$ relationship.

For further discussion, see Billington [1.5].

Example 1.3

Consider the previous example of a flat wall of surface temperature 20 °C with the air temperature of 10 °C. The heat loss determined by natural convection was 30 W/m². Consider now air forced over this wall at 4 m/s. What is the new heat loss per unit area?

For $v = 4\,\mathrm{m/s}$, h_c is given by

$$h_c = 5{\cdot}8 + 4{\cdot}1v = 22{\cdot}2\ \mathrm{W/m^2\ °C}$$

$$Q_{cv} = h_c A \, \Delta t$$

$$Q_{cv} = 22{\cdot}2 \times 1 \times 10$$
$$= 222\ \mathrm{W} \qquad \text{for } 1\ \mathrm{m}^2$$

i.e. the heat loss has been raised from 30 W/m² to approximately 220 W/m² by the forced air movement.

1.4.3 Applications of Convection

An immersion heater has the heating element placed in the lower regions of a domestic hot water tank. The heated water rises by natural convection to the top of the tank and is fed off at this point to the hot water taps, while the cold water inlet is at the bottom of the tank. The water tends to be stratified in the tank and the amount of hot water can simply be determined by feeling the side of the tank to find the hot/cold transition region.

More rapid cooking in ovens takes place at the top of the oven where the hot air has risen. However, forced convection is now often applied to cookers by fan assistance, so creating more even cooking conditions.

Hot water pipes will provide small amounts of heat to the surrounding air by natural convection. If the surface area of the pipes is increased by fitting vertical cooling fins, natural convection will be increased and if air is forced over the finned pipes with a fan, then a typical 'forced' or 'assisted convection' heater unit is produced with considerably increased heat transfer (see Section 5.3.2).

1.5 CONDUCTION

When the handle of a poker becomes hot because the end is in the glowing coals of a fire, conduction is being experienced. At the heated end the atoms of iron are vibrating more and more energetically and a transfer of this energy along the poker occurs, manifesting itself by a sensation of heat on the skin.

> *Conduction is the transfer of heat through a static material (usually a solid) by the atomic or molecular transfer of vibrational heat energy.*

Conduction is one example of the *flow* of a substance through a medium. The simplest example is the natural flow of water down a pipe. The amount of flow is measured by say the rate at which the water fills buckets at the end of the pipe. What governs this rate is firstly the driving force which in this case is gravity, being large if the pipe is steep, or small if the pipe is at a shallow angle. The second governing factor is the 'conductance' of the pipe, i.e. how easy it is for water to flow through the pipe. This will depend on the smoothness, cross-sectional area and straightness of the pipe.

In general the following relationship holds for a flow situation:

Amount of flow through a medium

= Driving force × Conductance of the medium

In the case of heat the amount of flow is the quantity of heat flowing in a given time, i.e. joules per second (J/s) or watts (W). The driving force is the temperature through which the heat is 'dropping' related to the distance over which this occurs. This is known as the temperature gradient, i.e. temperature difference divided by the distance. The conductance of the medium is a property which either makes it easy or difficult for heat to travel through it and is called the thermal conductivity.

1.5.1 Heat Flow Equation for Steady State

Consider a small section through a large plate of material as shown in Fig. 1.9.

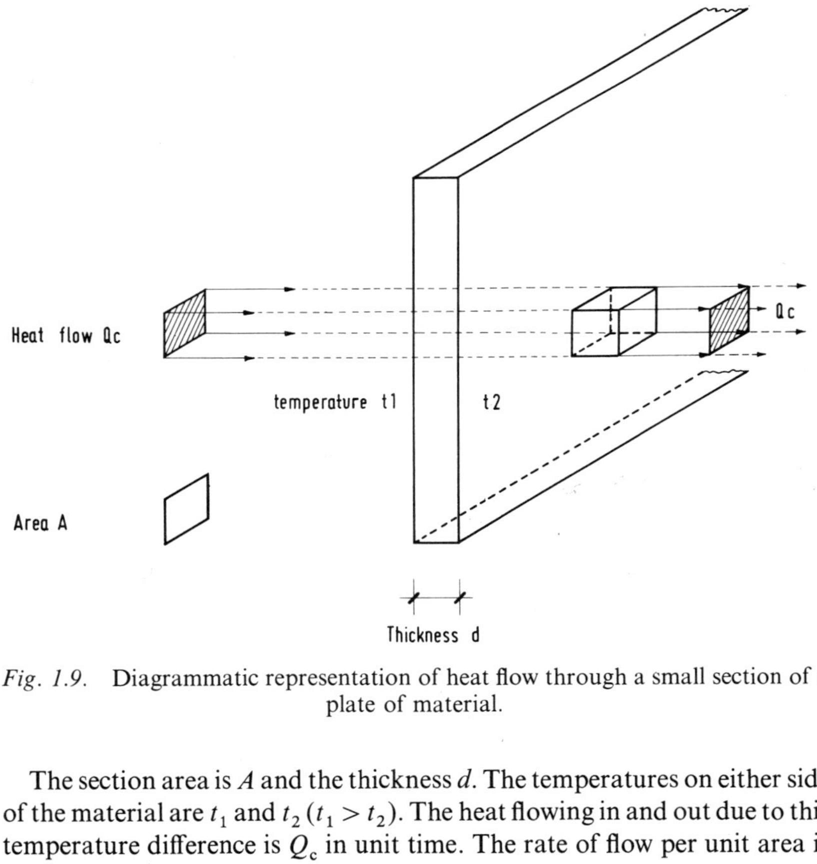

Fig. 1.9. Diagrammatic representation of heat flow through a small section of a plate of material.

The section area is A and the thickness d. The temperatures on either side of the material are t_1 and t_2 ($t_1 > t_2$). The heat flowing in and out due to this temperature difference is Q_c in unit time. The rate of flow per unit area is Q_c/A. 'Steady state' conditions are assumed to exist, i.e. the temperatures and heat flow are constant with time.

The driving force is the temperature gradient, as shown in Fig. 1.10.

The temperature gradient through the material along the line of heat flow

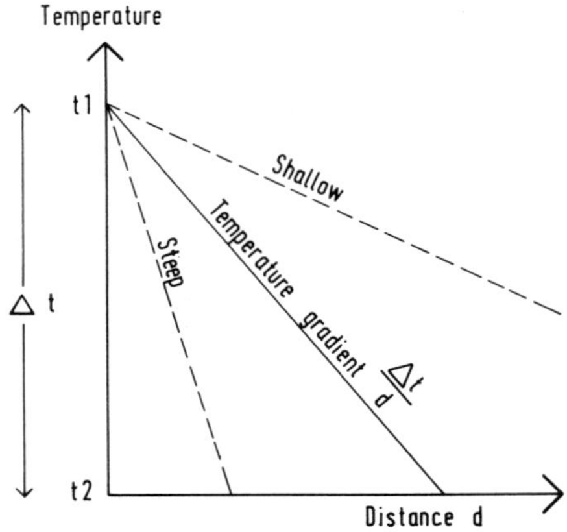

Fig. 1.10. Temperature gradients through a material.

is given by $(t_1 - t_2)/d$ or $\Delta t/d$. Let the conductance be k, the thermal conductivity. Then the flow equation

$$\text{Quantity} = \text{Driving force} \times \text{Conductance}$$

becomes

$$\frac{Q_c}{A} = \frac{\Delta t}{d} \times k$$

Rearranging gives the more usual expression of the standard conduction equation for steady state conditions:

$$Q_c = k \times A \times \frac{\Delta t}{d}$$

$$\underset{(\text{W})}{\text{Heat flow by conduction}} = \text{Thermal conductivity} \times \underset{(\text{m}^2)}{\text{Area}} \times \underset{(^\circ\text{C/m})}{\frac{\text{Temperature difference}}{\text{Distance}}}$$

1.5.2 Thermal Conductivity

If the above equation is rearranged, k is given by

$$k = Q_c \frac{1}{A} \frac{d}{\Delta t}$$

Making everything on the right hand side of the equation equal to unity except Q_c, then k, the thermal conductivity, can be defined as:

> *The thermal conductivity is the heat flow normal to a surface of unit area in unit time due to unit temperature gradient under steady state conditions.*

By substituting the units of each of the individual parameters, the units of k are seen to be W/m °C (W m^{-1} $°C^{-1}$), i.e. watts per metre degrees C.

Some typical values of k are given below (in W/m °C):

Copper (pure)	400
Steel	50
Brick	1·5–0·7
Timber	0·12–0·16
Fibre quilt	0·03–0·04
Still air	0·025

A common example of the effects of thermal conductivity is the difference in temperatures reached by metal and plastic handles of saucepans. The conductivity of the two materials differs by a factor of about 500, and the metal handle usually becomes too hot to touch when the pan is in use.

Conductivity values are closely related to the density of materials, as small pores or pockets of air in the material, which lead to low density, are also responsible for low conductivity because of the inherent low k value of air (0·025 W/m °C). The air is usually sufficiently well trapped to prevent localised heat transfer by convection. The high k values for metals are a result of more complex aspects related to their atomic structure. Values of thermal conductivity with increasing density for typical building materials are given in Table 1.2.

1.5.3 Thermal Resistivity and Resistance

It is often more convenient to think in terms of thermal resistivity rather than thermal conductivity. Conductivity is a measure of the *ease* with which heat can flow through a material, whereas resistivity is a measure of the *difficulty* of heat flow. Resistivity (r) is simply defined as the inverse of conductivity (k), i.e.

$$\text{Resistivity} = 1/\text{Conductivity}$$

$$r = 1/k$$

The reason for introducing the term is that for a series of materials linked

Table 1.2 Density and thermal conductivity values of some common building materials, based on [1.9]

Material	Bulk density (kg/m^3)	Thermal conductivity, k (W/m °C)
Asbestos cement	1 600	0·40
Asphalt	1 600–2 325	0·4–1·2
Brickwork:		
Common	1 600–2 000	0·6–1·0
Dense	2 400	1·6
Compressed straw slab	260	0·09
Concrete:		
Lightweight	450–1 000	0·10–0·30
Medium	1 500–1 700	0·4–0·7
Dense	2 000–2 400	1·0–1·8
Corkboard	145	0·04
Fibreboard	300	0·06
Glass	2 500	1·0
Glasswool, mat, quilt	25	0·04
Hardboard:		
Medium	600	0·08
Standard	900	0·13
Metals:		
Aluminium alloy, typical	2 800	160
Copper, commercial	8 900	200
Steel	7 800	50
Mineral wool quilt	50	0·04
Plaster:		
Lightweight	400–900	0·08–0·3
Dense	1 600	0·53
Plasterboard	960	0·16
Plastics:		
Cellular:		
Expanded polystyrene	15–30	0·03–0·04
Urea formaldehyde	8	0·04
Polyurethane	30	0·03
Solid:		
Epoxy glass fibre	1 500	0·23
Polystyrene	1 050	0·17
Plywood	530	0·14
Stone	2 000–2 800	1·3–2·9
Timber	640	0·14
Wood chipboard	800	0·15
Woodwool slab	500–800	0·1–0·15

Values are for typical moisture contents and temperatures.

together resistivity effects are additive, which is not the case with conductivity.

The thermal resistance (R) brings in the thickness (d) of the material and is defined as

$$\text{Resistance} = \frac{\text{Thickness}}{\text{Conductivity}} \quad \text{or} \quad \text{Thickness} \times \text{Resistivity}$$

i.e.

$$R = \frac{d}{k} \quad \text{or} \quad dr$$

In other words the greater the thickness of material, the greater is its resistance.

Establishing the overall resistance of a series of materials combined together is simply a matter of summing their individual resistances, i.e.

$$R_{\text{Total}} = R_1 + R_2 + R_3 + R_4 \ldots \qquad \text{i.e.} \sum R_i$$

This process is incorporated in the calculation of U-values, which follows.

1.6 THERMAL TRANSMITTANCE OR *U*-VALUE

For applications in buildings the flow of heat through a building element from the space on one side to that on the other side is of interest. The parameter which gives a measure of the conductance of a particular constructional element for these circumstances is the thermal transmittance or U-value. The latter term comes from the old 'British Thermal Unit' of heat. The thermal transmittance takes into account the inherent thermal conductivities and thicknesses of the materials involved together with the effects of surfaces on the flow of heat. This latter topic must be considered before the application and calculation of U-values is described.

1.6.1 Surface Resistance

Heat is transferred from a solid surface to the surroundings or from the surroundings to a solid surface by radiation and convection processes. The radiation transfer process is dependent on the surface absorptivity or emissivity and on the radiation temperatures involved, as described in Section 1.3.4. Thus, for instance, the radiation loss from polished aluminium at ambient temperatures ($\varepsilon \sim 0{\cdot}1$) is much less than from brick ($\varepsilon \sim 0{\cdot}9$).

The convection transfer process at a surface depends on the nature of the boundary layer of air which tends to be held by the surface giving the effect of a thin blanket. The factors involved, described in Section 1.4, are the air and surface temperature differences, the orientation of the surface and the local velocity of air. A further factor is the surface roughness. A molecular rough surface will tend to trap and hold the blanket of air more readily than an optically rough surface which tends to act as a set of cooling fins assisting heat transfer. Surface roughness can affect the resistance to convection heat transfer by up to 30% (e.g. [1.6]).

The heat transfer equations for radiation and convection, derived in Sections 1.3.4 and 1.4.1 respectively, can be applied to the effects occurring at the surface. The convection coefficient, h_c, is a measure of the ease with which heat is transferred by convection, while the emissivity × the radiation coefficient, εh_r, is a measure of the ease of radiation transfer. As mentioned previously, the values of the coefficients depend on the temperature differences involved.

Combining the two effects it can be shown that the conductance of the surface is given by the term $(h_c + \varepsilon h_r)$, i.e. the sum of the two conductances [1.5, 1.7]. More usually the resistance is quoted, i.e. $1/(h_c + \varepsilon h_r)$.

For normal temperature differences at internal vertical walls the usual values for the parameters are

$$h_c = 3{\cdot}0\ \mathrm{W/m^2\ °C}$$
$$h_r = 5{\cdot}7\ \mathrm{W/m^2\ °C}$$
$$\varepsilon = 0{\cdot}9 \text{ (for ordinary building materials)}$$

Substituting these gives a surface resistance value of $0{\cdot}123\ \mathrm{m^2\ °C/W}$.

Similarly, substitutions can be made for any surface, exposure and orientation, giving for instance a typical resistance for an external vertical wall of $0{\cdot}06\ \mathrm{m^2\ °C/W}$. Such a value relates to average conditions of wind and temperature and should always, therefore, be considered as an approximation for any new situation.

Generally speaking for external surfaces the temperature differences and air velocities are higher and surfaces are rougher, giving lower resistance values. Further details of factors which in practice affect surface resistances are given in Section 3.2.2.

It is of interest to note that the total resistance of a sheet of window glass is virtually completely dependent on the two surface resistances, the resistivity of the glass itself providing very little contribution

(Section 3.2.4). Similarly the resistance of an air cavity which is sufficiently wide to allow restrained convection currents to circulate (i.e. 20 mm and greater) is essentially a combination of the two surface resistances involved with a value typically of 0·18 m^2 °C/W (Section 3.2.5).

1.6.2 Environmental Temperature

The flow of heat into or from a surface depends on both convection and radiation transfer, and hence on both the air temperature and mean radiant temperature as seen by the surface. It is of convenience for heat transfer calculations' purposes to combine the air and radiant temperatures into a single temperature value, known as the environmental temperature.

> *The environmental temperature is the temperature of a hypothetical uniform environment (with air and radiant temperatures equal) which would give the same rate of heat transfer through a building element as occurs under the prevailing conditions* [1.8].

It can be shown [1.7] that the appropriate weighting of the air temperature and mean radiant temperature to satisfy this definition for a cubic room is given by

$$\text{Environmental temperature} = \tfrac{2}{3}(\text{Mean radiant temperature}) + \tfrac{1}{3}(\text{Air temperature})$$

This proportioning is a reflection of the relative rates of heat transfer by radiation and convection. It was shown in the examples in Sections 1.3.4 and 1.4.1 using simplified expressions for heat transfer that for an object at 20 °C in surroundings of air and mean radiant temperature of 10 °C the rate of heat loss was 50 W/m^2 for radiation and 30 W/m^2 by convection. The proportions are close to the two-thirds, one-third as defined by the environmental temperature.

Thus it is not simply the air temperature adjacent to a wall which affects the flow of heat to or from the wall, but rather a combination of the air and mean radiant temperature. Hence, this should be taken into consideration in the calculation of fabric heat transfer.

1.6.3 Fabric Heat Flow

When considering heat flow through a building fabric from one space to another the form of the original conduction heat flow equation applies but the terms are adapted to fit the new circumstances. The flow equation, derived in Section 1.5.1, is

$$\begin{matrix} \text{Heat flow} \\ \text{by conduction} \end{matrix} = \begin{matrix} \text{Thermal} \\ \text{conductivity} \end{matrix} \times \text{Area} \times \frac{\text{Temperature difference}}{\text{Distance}}$$

$$Q_c = k \times A \times \frac{\Delta t}{d}$$

The terms are revised in the following fashion:

(a) The heat flow by conduction becomes the heat flow through the fabric, Q_f, i.e. from the space on one side of the fabric to the space on the other side.
(b) The thermal conductivity value, which is specified for one material, becomes the thermal transmittance or U-value, which is the conductance of a total component and may be of many materials, and includes the conductances of surfaces involved.
(c) The temperatures used should be the environmental temperature, t_e, as it has been shown to be the appropriate temperature parameter when considering heat flow into or from a surface.
(d) The distance component is absorbed into the U-value. Thus the new expression is

$$\begin{matrix} \text{Heat flow} \\ \text{through fabric} \end{matrix} = U\text{-value} \times \text{Area} \times \begin{matrix} \text{Environmental temperature} \\ \text{difference across fabric} \end{matrix}$$

i.e.

$$Q_f = U \times A \times \Delta t_e$$

The means of calculating U-values is given in the following section, while the application of the fabric heat transfer equation is demonstrated in Section 3.3.

1.6.4 Calculation of U-values

The U-value of a constructional element is determined by summing the total resistance of the materials and surfaces involved and then inverting this resistance to give a transmittance or conductivity value (e.g. [1.9]).

Thus

$$U\text{-value} = 1 \Bigg/ \left(\begin{matrix} \text{Internal} \\ \text{surface} \\ \text{resistance} \end{matrix} + \begin{matrix} \text{Resistance} \\ \text{of material} \\ 1 \end{matrix} + \begin{matrix} \text{Resistance} \\ \text{of material} \\ 2 \end{matrix} + \cdots + \begin{matrix} \text{External} \\ \text{surface} \\ \text{resistance} \end{matrix} \right)$$

$$= 1/\text{Total resistance, i.e. } 1/R_{\text{Total}}$$

The resistance of each component material is given by the thickness (in metres) divided by the conductivity (k) value involved (Section 1.5.3), while the resistances of surfaces or of any air space in the construction are usually taken as quoted values (e.g. from Section 1.6.1 or Section 3.2).

Example 1.4

What is the U-value of a wall of the following construction: 105 mm outer brick leaf (medium density), 50 mm cavity, 100 mm inner lightweight (aerated) concrete block leaf and 15 mm plaster (dense)?

k-values are taken from Table 1.2.

Material	Thickness (m)	k-value (W/m °C)	Resistance (m^2 °C/W)	{(Thickness/k-value) or as quoted for surfaces and air spaces}
External surface			0·06	
Brick	0·105	0·94	0·112	
Cavity			0·18	
Lightweight concrete	0·100	0·18	0·556	
Plaster	0·015	0·53	0·028	
Internal surface			0·12	
	Total resistance		1·056	

$$U\text{-value} = \frac{1}{\text{Total resistance}} = \frac{1}{1{\cdot}056} = 0{\cdot}947 \text{ W/m}^2 \text{ °C}$$

There is not normally sufficient confidence in the basic conductivity and surface resistance data due to the variability caused by effects such as density, moisture and exposure (see Section 3.2) to quote U-values more accurately than to about 10%. Therefore values greater than 1 are usually quoted to one decimal place and lower values to two decimal places. In this example the value would usually be approximated to 1·0 W/m^2 °C. Further discussion of U-values occurs in Section 3.2.

1.7 HEAT TRANSFER BY CHANGES OF STATE

1.7.1 Melting and Evaporation Processes

When the heat content of a solid material is increased the atoms or molecules eventually vibrate so actively that they break away from each other forming a liquid material (i.e. a material where the bonding between atoms and molecules is very loose). If further heat energy is added then this liquid will convert into the gaseous state, where the heat energy of all the atoms or molecules is sufficient to make them move freely in space. Thus the sequence

$$\text{Solid} \rightarrow \text{Liquid} \rightarrow \text{Gas}$$

occurs with increasing heat content, or a rising of temperature, and the reverse process will occur on cooling down.

There are special cases where the liquid phase is bypassed. This occurs with certain chemicals, but a common example is the disappearance of frost from the ground without its passing through the liquid phase.

Whether a material exists under normal temperature conditions as a gas, liquid or solid depends on the inherent binding forces in the material, compared to the heat content causing movement.

The change of state from a solid to a liquid, or from a liquid to a gas, requires an additional quantity of heat to break down the binding forces existing at the surface of the material. This additional heat requirement is known as the latent heat of fusion for solids to liquids and the latent heat of evaporation for liquids to gases.

1 kg of ice at 0 °C converted to 1 kg of water at 0 °C requires a heat input of 333 kJ, while 1 kg of water at 100 °C requires 2257 kJ to convert it to 1 kg of steam. If the reverse processes occur then these quantities of heat are released.

The process of evaporation of moisture from a surface is the most common example of a change of state leading to heat transfer. For instance, if 1 g of water (perspiration) is evaporated off the surface of the skin then more than 2000 J of heat are lost from the body into the atmosphere through the latent heat of vaporisation. The principle is commonly applied in hot countries for keeping water flasks cool; the flask is wrapped in a moist cloth (or becomes moist, being permeable animal skin or porous clay) and the evaporation at the outside surface creates cooling. The process of evaporation of a liquid to create cooling is also applied in the common refrigerator.

Cooling by evaporation is of great importance when considering the thermal performance of the human body (Section 2.2). Liquid/vapour transitions are also relevant in considering moisture in buildings and the wide range of effects relating to relative humidity, moisture content and condensation. For these topics to be understood and applied the subject of psychrometrics, i.e. the moisture content properties of the air, must be examined.

1.7.2 Psychrometrics

A given mass of air at a fixed temperature is able to absorb a certain amount of water vapour. It can hold any amount up to a given maximum at which point the air is termed 'saturated'. If the mass is then heated up it is able to absorb more vapour before it becomes saturated again. On the other hand a mass of saturated air on being cooled must lose some of the water vapour it holds, and this converts to liquid water, the process being called condensation.

The amount of water vapour in the air can be expressed by the *water content* or *absolute humidity* which are measures of the mass of water vapour in a given mass of air (i.e. kilograms of water vapour in a kilogram of air). A typical value under normal conditions is 0·01 kg of water held in 1 kg of air (which occupies approximately 0·85 m^3).

Saturated conditions at a given temperature are similarly expressed. The maximum vapour content of the air increases with temperature as demonstrated in Fig. 1.11, which plots the saturated absolute humidity for a given mass of air against air temperature.

A pressure results from the water vapour in the air, known as the *vapour pressure*. This contributes a small amount to the total atmospheric pressure (e.g. a few millibars of vapour pressure in 1000 millibars of atmospheric pressure). The water vapour pressure increases with the water vapour

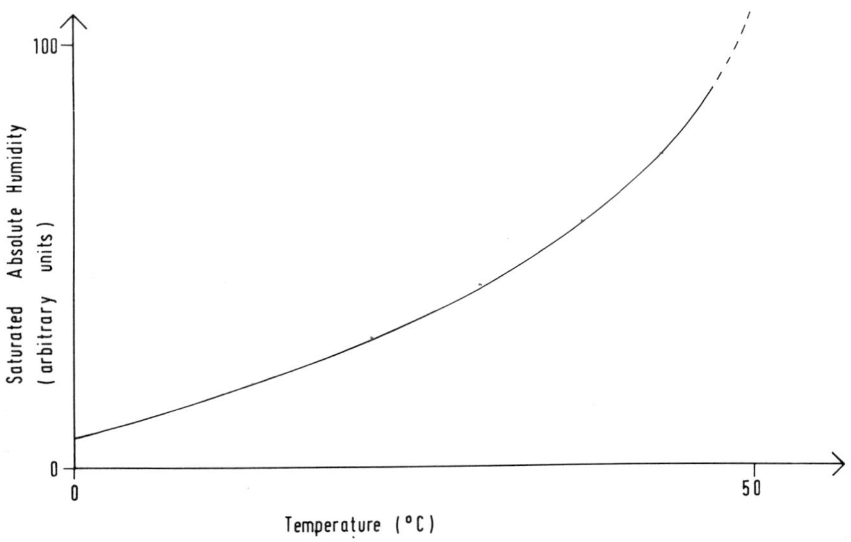

Fig. 1.11. Saturated absolute humidity (i.e. maximum water vapour content) versus temperature of the air.

content of the air, i.e. the absolute humidity. (The relationship is approximately linear.) The vapour pressure is responsible for driving water vapour from high to low regions of absolute humidity which can result in condensation within materials or constructions, i.e. interstitial condensation. This is further explained in Section 3.7.

A more convenient method of describing the moisture content of air is by the term *relative humidity* (RH) which expresses the amount of water vapour in the air (in terms of its vapour pressure) as a percentage of the total possible (i.e. the saturated value) in the air at the same temperature and atmospheric pressure. Thus 50 % relative humidity means that only half of the possible water vapour (based on its pressure) is held in the air while 100 % RH means that no more water vapour can be held, i.e. saturated conditions occur.

A similar term to relative humidity is also used which is *percentage saturation*. This again expresses the water vapour in the air as a percentage of that which is possible for given conditions of temperature and pressure. However, in this case the vapour is measured in terms of the mass of water vapour present rather than the vapour pressure it exerts. As previously mentioned, the vapour pressure and mass of vapour in the air are closely related, so that the values of relative humidity and percentage saturation for a particular set of conditions are usually very close.

The relationships between the air temperature (also termed the dry bulb temperature), the water content or absolute humidity (and associated vapour pressure) and relative humidity are summarised in the psychrometric chart (Fig. 1.12). The air temperature is selected on the base line and the absolute humidity relating to a given relative humidity value can be read off.

A further parameter introduced on the psychrometric chart is the *wet bulb temperature*. This is the temperature that a thermometer will take up if it is moist (e.g. a wet wick is around it) and the moisture is allowed to evaporate freely. A cooling effect occurs which depends on the amount of evaporation occurring. This in turn depends on the amount of moisture vapour the air can accept at the given air or dry bulb temperature. The lower the relative humidity the more evaporation is possible and hence the lower the wet bulb temperature. Alternatively at 100 % relative humidity no evaporation can take place and the wet and dry bulb values are identical.

There are three major applications of the wet bulb temperature. The first is a means of estimating the amount of cooling that will take place when moisture is evaporating from a surface (e.g. a human body). The second is that it provides a simple means of measuring relative humidity. For example, the simple whirling hygrometer has two thermometers which are

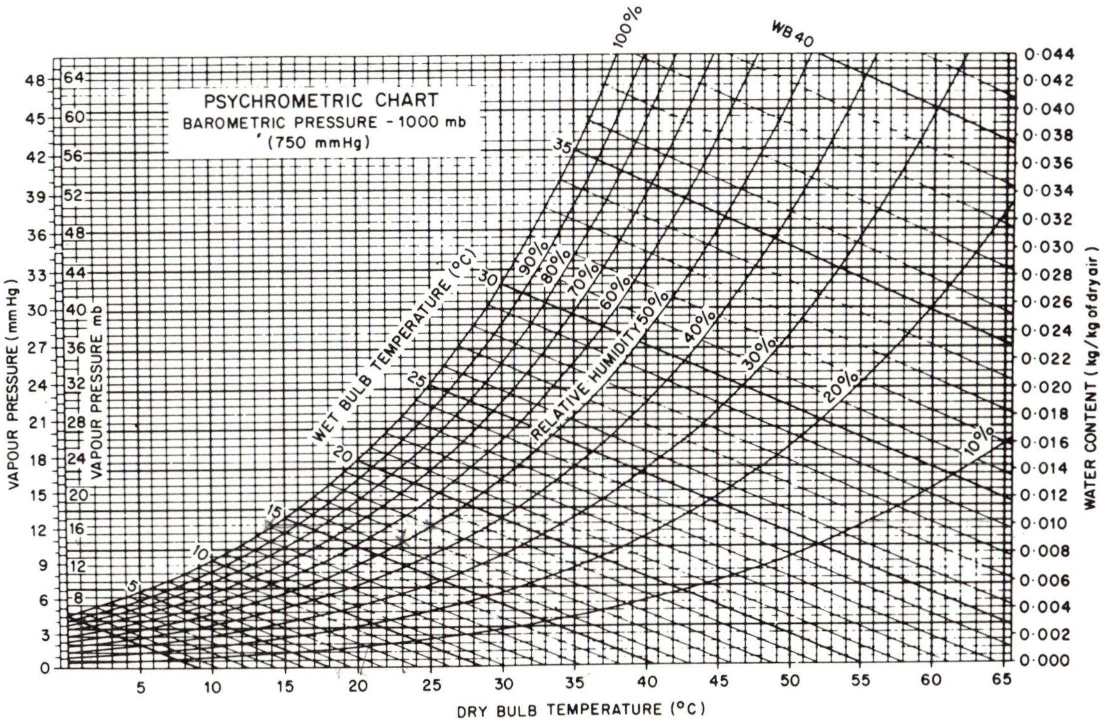

Fig. 1.12. The psychrometric chart showing the relationship between dry and wet bulb temperature and relative and absolute humidities.

whirled to provide consistent evaporation conditions. One thermometer measures the dry bulb temperature; the other has a moist wick over its bulb, measuring the wet bulb temperature. The relative humidity may then be obtained from the psychrometric chart or from tables which relate the two temperatures.

The third application is in the prediction of condensation conditions. If the dry bulb and wet bulb values are known then the temperature at which condensation occurs can be read off by following a horizontal line on the chart (i.e. keeping absolute humidity constant) until the 100 % RH value is reached, i.e. the wet bulb temperature at 100 % RH is given. This is then the temperature at which condensation occurs, i.e. the dew point.

Example 1.5

The dry bulb is measured at 25 °C and the wet bulb at 18 °C. What is the relative humidity and at what temperature will condensation occur?

Referring to the psychrometric chart, Fig. 1.12, then the intersect of wet and dry bulb values gives an RH value of 50 %. Reading horizontally from the intersect point, i.e. keeping absolute humidity constant (at 0·01 kg/kg) then 100 % saturation is at 14 °C, i.e. condensation occurs at 14 °C or below. The assessment of whether a surface will be at this temperature is given below.

1.7.3 Prediction of Surface Temperatures

To predict surface temperatures in a room requires knowledge of the *U*-value of the construction involved and likely inside and outside temperatures. For an external wall and cold outside conditions a temperature drop will occur inside the room from the air to the surface of the wall, because of the surface resistance. The drop that occurs, and hence the surface temperature, can be calculated for steady conditions of temperature. (Strictly speaking the temperatures on either side of the wall are the environmental temperatures, but these are assumed to be equal to the air temperatures for simplicity in the analysis.) The ratio of the surface temperature drop to the total temperature drop from inside to outside is the same as the surface resistance to the total resistance of the construction. Thus if the internal surface represents a large proportion of the total resistance (e.g. single glazing) the temperature drop will be high and the surface will be cold compared to inside the room. However, for low *U*-value constructions, the internal surface resistance represents a small proportion of the total resistance and the temperature drop is small. The full procedure is described elsewhere (e.g. [1.10]).

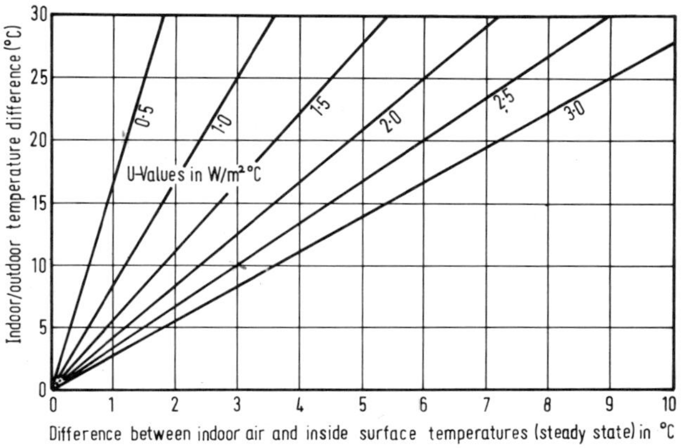

Fig. 1.13. Plot enabling the surface temperature of a construction of a given U-value to be determined for a given temperature drop from inside to outside for steady temperature conditions (adapted from [1.10]).

The information can be summarised graphically as shown in Fig. 1.13. For a given temperature difference from inside to outside, and a known U-value of construction, the inside surface temperature drop at the material can be read off. Thus, for instance, for an inside temperature of 18 °C and outside −2 °C (temperature drop 20 °C), and a U-value of 1·0 W/m^2 °C, the temperature drop at the surface will be 2·5 °C, and the surface temperature will be (18 − 2·5), i.e. 15·5 °C.

For single glazing (U-value approximately 5·0 W/m^2 °C) in these circumstances the surface temperature inside will be 6 °C. This is determined from Fig. 1.13 by doubling the temperature drop occurring for a U-value of 2·5 W/m^2 °C. Thus referring back to the psychrometric chart, condensation would occur on the window for relative humidities in the room as low as 46 %.

Further discussions of condensation problems relating to thermal conditions in dwellings and to thermal insulation materials are given in Sections 2.4 and 3.6. The overall topic has also been extensively documented elsewhere (e.g. [1.10, 1.11, 1.12]).

1.8 THERMAL CAPACITY AND THERMAL MASS

A further important thermal characteristic of building materials relates to the amount of heat the material can 'store' or alternatively the amount of heat required to raise the temperature of the material by a given amount.

The property which governs this characteristic is the *thermal capacity* (or specific heat) of the material.

The thermal capacity is the quantity of heat required to raise 1 kg of the material through 1 °C.

The units are J/kg °C. Typical values of thermal capacity are given in Table 1.3. From this table it is apparent that thermal capacity values do not vary a great deal for common building materials, except that water has a particularly high value. This is fortuitous when considering water as a medium to transfer heat around buildings through 'wet' systems.

Table 1.3 Values of thermal capacity, density and volumetric thermal capacity for some building materials, based on [1.5]

Material	Approximate thermal capacity (c) (J/kg °C)	Density (ρ) (kg/m³)	Typical volumetric thermal capacity ($\rho \times c$) (J/m³ °C) $\times 10^{-3}$
Concrete	950	800–2 200	700–2 100
Brick	1 000	1 700–2 300	1 700–2 300
Timber	1 500	500–1 200	750–1 800
Insulating fibreboard	1 400	300	400
Steel	500	7 800	3 900
Granite	900	2 700	2 400
Water	4 200	1 000	4 200
Air	1 100	1·2	1·3

The thermal capacity is defined for unit *weight* of material. However, it is sometimes more appropriate to consider the capacity of a *volume* of material, particularly when considering storage effects combined with conductive heat flow into or out of a material where the distances involved affect the conduction process. Such considerations lead to the parameters known as thermal diffusivity and thermal penetration coefficient, which are described in Section 5.2. In such cases the *volumetric thermal capacity* is of relevance. This is given by:

$$\text{Volumetric thermal capacity (J/m}^3\ {}^\circ\text{C)} = \text{Thermal capacity (J/kg}\ {}^\circ\text{C)} \times \text{Density (kg/m}^3\text{)}$$

and values are given in Table 1.3. It is apparent that there are relatively large differences in the volumetric thermal capacities quoted, but these arise essentially from the wide range of densities of the materials involved.

When considering a building or construction as a whole then the thermal storage capacity is termed the *thermal mass*. This takes into account the total weight of the materials involved and is given by

Thermal mass (J/°C)

= Thermal capacity (J/kg °C) × Weight of material (kg)

Thus, for example, the air in a room has a low thermal mass despite the relatively high thermal capacity value (Table 1.3) because the weight of the material is low. In a room of dimensions 4 m × 3 m × 2·4 m the weight of air is approximately 35 kg, which is the weight of about ten standard bricks. Similarly insulation materials have low thermal mass, whereas granite and steel in a building usually have a high thermal mass because the weight of material involved is high.

The thermal mass of a building is hence related to the weight of the materials involved and the terms heavyweight and lightweight are used in this context, being related to the quantities and densities of materials used. Thus a concrete structure with brick walls represents a 'heavyweight' structure physically and thermally; it stores a large amount of heat, but also requires large heat inputs to warm it up. A timber frame building with timber cladding is a 'lightweight' building and the opposite applies. However, the parallel must be used with care as, for instance, a heavyweight building can be made to respond internally like a lightweight building by positioning insulation internally (see Section 5.2.3).

One effect of thermal mass in thermal design involves the variation of the temperature in the building as external temperature conditions change. A heavyweight building is much slower to respond to external fluctuations than a lightweight one. This is best illustrated by buildings in hot climates which even out the very high temperatures (from sun and air) during the day and relatively low temperatures at night. A typical dwelling in a Mediterranean country is constructed of thick stone walls, or concrete and brick. These materials steadily heat up during the day and dissipate their heat at night. A lightweight construction would follow much more closely the fluctuations of the external conditions.

These effects are discussed and illustrated in Section 5.2, together with the effect of the positioning of insulation on the thermal mass characteristics. Other parameters, such as thermal diffusivity, which are helpful in describing the effects of changing thermal conditions, rather than 'steady state' conditions, are also introduced in this section.

REFERENCES

1.1 FANGER, P. O.: *Thermal Comfort: Analysis and Applications in Environmental Engineering*. McGraw–Hill, New York, 1970.

1.2 CHRENKO, F. A. (Ed.): *Bedford's Basic Principles of Ventilation and Heating*. H. K. Lewis, London, 1974.

1.3 TURNER, D. P. (Ed.): *Windows and Environment*. Pilkington Brothers, St. Helens, 1969.

1.4 MCGRATH, R., FROST, A. C. and BECKETT, H. E.: *Glass in Architecture and Decoration*. The Architectural Press, London, 2nd edn, 1961.

1.5 BILLINGTON, N. S.: *Building Physics: Heat*. Pergamon Press, London, 1967.

1.6 GIVONI, B.: *Man, Climate and Architecture*. Applied Science, London, 2nd edn, 1976.

1.7 DANTER, E.: Heat exchanges in a room and the definition of room temperature. *J. Inst. Heat. Vent. Eng.*, Vol. 41 (Feb. 1974), pp. 232–245.

1.8 CHARTERED INSTITUTION OF BUILDING SERVICES: *CIBS Guide A5: Thermal Response of Buildings*. CIBS Publications, London, 1979.

1.9 BUILDING RESEARCH ESTABLISHMENT: *Digest 108: Standard U-values*. HMSO, London, 1975.

1.10 BUILDING RESEARCH ESTABLISHMENT: *Digest 110: Condensation*. HMSO, London, 1972.

1.11 MARKUS, T. A. and MORRIS, E. N.: *Buildings, Climate and Energy*. Pitman, London, 1980.

1.12 CROOME, D. J. and SHERRATT, A. F. C.: *Condensation in Buildings*. Applied Science, London, 1972.

Chapter 2

Thermal Comfort and Air Quality

2.1 INTRODUCTION

One of the fundamental reasons for humans providing shelter for themselves is for protection against variations or extremes of climate. The earliest civilisations developed in hot countries and required protection from direct sun and low temperatures at night. Simple huts provided primitive protection whilst cave dwellings would also even out daily and seasonal temperature fluctuations. As civilisation developed in cooler climates buildings became of paramount importance for protection against cold, wind and rain. At the same time buildings helped to contain heat which was being generated by inhabitants and by the burning of fuels. No great emphasis was placed on insulation values of the fabric but use was often made of local climatic conditions by the appropriate orientation and form of the building (e.g. [2.1]).

In the twentieth century, buildings have developed to the extent where full artificial control of the internal conditions through both heating and cooling can be achieved, often with complete disregard for prevailing external conditions (e.g. [2.2]).

The ability to fully control internal environments has made it possible to generate within buildings the 'ideal' thermal conditions for people to live and undertake activities. In the past people worked out the conditions best suited for themselves through instinct and sensation, and the creation of such conditions within buildings was invariably achieved on a rather crudely controlled basis (although not negating success). Nowadays, most buildings provide heating or cooling so that occupants can generate 'comfort' conditions while, particularly in large buildings, conditions tend to be specified for the occupants and maintained by the mechanical services (in conjunction with the building fabric) with little or no facility for personal control.

The provision of thermal inputs, and in many cases the facility for full mechanical control of the thermal environment, is an initial responsibility and function of the building design team. The team must not only be able to decide what conditions need to be achieved for the particular activities involved in the building, but also appreciate what problems might arise for the occupants, fabric and contents if deviations occur from these conditions.

The design team must also establish the best ways of achieving these conditions through both the manipulation of building form and fabric and through the appropriate selection and application of mechanical services. At the same time a major concern is to create these conditions as economically as possible in both capital and running cost terms.

This chapter examines the basic requirements for the thermal conditions necessary for people to live and perform comfortably, healthily and effectively, and later chapters examine the means by which these conditions are achieved.

2.2 THERMAL PERFORMANCE OF THE HUMAN BODY

Human beings are warm blooded animals. This means that their body core temperature (i.e. the temperature deep in the body tissues) must be regulated to remain within a narrow range, namely 36·9 ± 0·5 °C (98·4 ± 1 °F); the process is known as heat homeostasis. The reason for this is that many of the critical biochemical and cellular processes on which bodily functions depend take place efficiently and correctly only within this narrow range of temperatures. If the core temperature drops to about 31 °C consciousness is lost and death can ensue rapidly; if the temperature rises to about 43 °C, thermoregulation is lost, and unconsciousness and death can again ensue. Hence, for simple health and survival reasons, this constancy of temperature must be maintained (see, for example, Newburgh [2.3]).

The human body extracts energy from its surroundings by consuming food to generate an ordered self-contained system, i.e. the body as a whole, which can function efficiently to maintain the energy input on which it depends. The input is required to sustain cellular functions, produce new cells and provide muscular action. This energy is all created by the breakdown of food, with an associated generation of thermal energy. A typical individual may consume 12 MJ of energy in a day, and the majority of this (70–80 %) is converted to heat. This heat contributes to maintaining body temperature and when the body is performing normally there is excess for this requirement and heat must therefore be lost to the surroundings. The climates in which humans survive successfully, without excessive dependence on technology, are those providing temperature conditions which enable the heat loss process of the body to take place effectively and in a controlled fashion.

The rate of heat generation, i.e. the metabolic rate, and hence of heat loss is closely related to the nature of the activity that the body is undertaking. Typical figures for a range of activities are given in Table 2.1.

Table 2.1 Typical metabolic rates for varying activities (adapted from Fanger [2.4])

Activity	Metabolic rate for a typical adult (W)
Sleeping	75
Sitting	105
Standing	125
Walking at 3·2 km/h	210
Walking at 6·4 km/h	270
Heavy work	360–450

Thus the greater the activity the higher the rate of heat loss required to keep the body temperature constant, while at low activities the usual problem is to reduce heat loss sufficiently to avoid a fall in body temperature. The mechanisms by which the body gains and loses heat are now considered.

2.2.1 Heat Transfer To and From the Body

The four modes of transfer of heat—conduction, convection, radiation and evaporation—all contribute towards the thermal balance of the human body.

Conduction plays a minor role in that it essentially only operates when the body is in direct contact with solid objects, e.g. bare feet on a cold floor, or a bare arm on the metal rest of a chair. If the loss or gain of heat by the contact is high this is immediately sensed and remedial action taken. Conductivity values, however, play an important role in terms of the insulation value of clothing.

Radiation to and from the body depends on the surface temperature of the skin and the radiant temperature of the surroundings. Usually the net result is radiation loss to the surroundings, but, for instance, lying in the sun can produce a net gain.

Convection losses depend on the air temperature and air movement. A windy day 'feels' colder than a still day, even if the air temperatures are identical, because of the forced convection effect. However, sitting in front of a warm air fan heater will cause localised heat gain by convection because the air temperature is higher than the body surface temperature.

Evaporation losses, occurring through respiration and perspiration, depend on the relative humidity, temperature and movement of the air and the rate at which moisture is being produced. This mechanism always contributes to heat loss, not to heat gain, and becomes the major

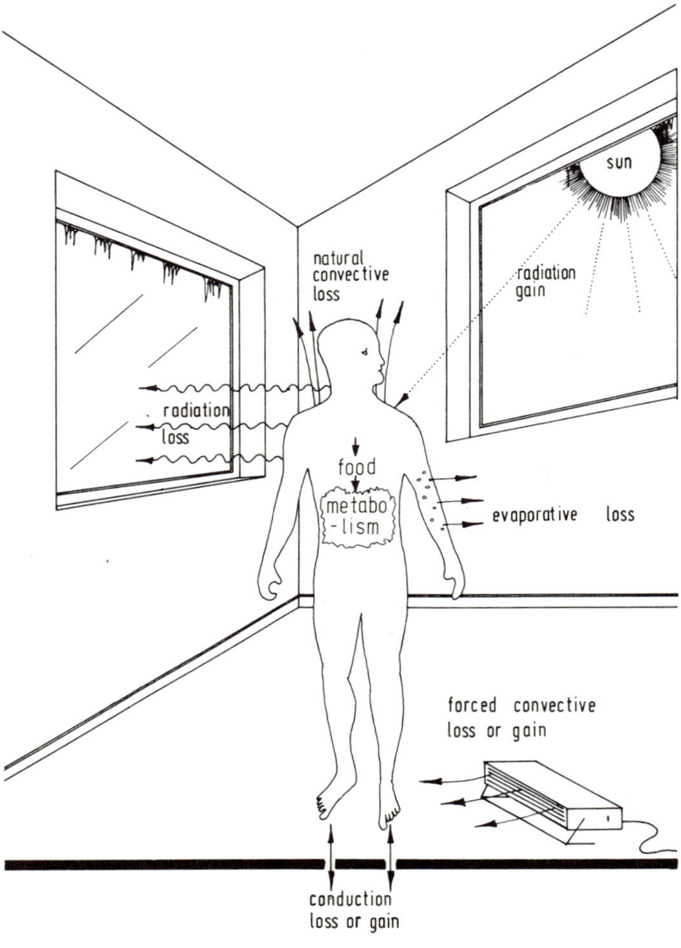

Fig. 2.1. The human body: heat losses and gains.

mechanism for loss when air temperatures exceed about 25 °C. The four mechanisms are summarised in Fig. 2.1.

For heat balance to be maintained the heat generated internally combined with any changes in the heat stored in the tissues must be balanced by these four heat loss (or gain) mechanisms. The balance is maintained to some extent by the control of heat production, but mainly by the regulation of heat loss, and is summarised below.

$$M \pm S = E \pm R \pm C \pm Cd$$

Body heat production (metabolic rate minus energy expended in working)	±	Changes in heat stored in tissues	=	Evaporative loss	±	Radiation exchange	±	Convection exchange	±	Conduction exchange

A number of workers have attempted to quantify the proportion of heat lost by convection, radiation and evaporation under typical comfort conditions. (The proportion lost directly by conduction is normally insignificant.) These proportions are quoted typically as 45 % radiation, 30 % convection and 25 % evaporation (e.g. [2.5]), but they vary considerably with changing external factors. For instance, cold, windy weather will cause the convection to rise, while in hot climates the evaporation component becomes dominant (see following section).

2.2.2 *The Thermoregulatory System*

The temperature of the bloodstream flowing in the core tissues is sensed in the regulation centre at the base of the brain known as the hypothalamus (which is responsible for a number of other sensing and control processes). If the blood temperature is sensed to be too high then the blood is made to flow closer to the outside surface of the body by an 'opening-up' of the blood vessels (vaso-dilation) and more heat is lost by convection and radiation because of the increased body surface temperature. If the blood temperature is sensed to be too low a 'closing-down' of blood vessels occurs (vaso-constriction) and heat loss is reduced. There are also hot and cold sensors in the skin which cause the same effects by reflex action on a localised basis. The shift of blood flow explains why people appear red and flushed when they are hot and white when cold.

The temperature sensors in the skin are also responsible for reflex muscular action when high or low temperature surfaces are touched; this is an essential protection mechanism to avoid damage to tissues. In cold conditions other reflex actions cause hairs on the skin to become erect, hence trapping more air close to the skin to enhance the insulation blanket. Also shivering occurs; this is a rapid contraction and expansion of peripheral muscles which has the effect of producing heat rapidly to warm up the outer surface of the body.

Moisture is continuously evaporated through the skin (known as insensible perspiration) and from the lungs, and this accounts for about 25 % of heat lost under typical comfort conditions. This process is not controlled and the evaporative loss depends on the psychrometric properties of the surrounding air. However, when an increase in heat loss from the body is required (either because of activity or raised external temperatures) sensible perspiration, i.e. sweating, occurs from sweat glands in the skin, the sweat production rate depending on the conditions, and more heat is lost by evaporation. This form of temperature control is well developed in man and indicates that he was originally a tropical animal, for under high external temperature conditions the only form of heat loss is by sweat evaporation.

In typical comfort conditions less than 0·1 litres of moisture is lost from the body per hour. This rises to approximately 0·5 litres per hour for light work or in warm temperatures (about 28 °C), but may rise to 1 to 2 litres per hour or more for heavy work and high temperatures (35 °C) [2.6]. If high sweat rates continue for an extended period then the body loses large amounts of water and salt and if these are not replenished in a regular fashion then medical effects occur such as heat exhaustion, heat cramp and thermal anhydrosis (insufficient fluid for full sweating to take place) [2.6]. The body's core temperature will also tend to rise above the norm because heat loss is insufficient. In the extreme case this leads to heat stroke, which is a breakdown of the thermoregulatory system followed by an uncontrolled rise in the body core temperature, unconsciousness and eventually death unless remedial action is taken rapidly. This involves a rapid artificial cooling of the body which can be achieved by spraying the body with water and inducing convection and evaporation by forced air movement.

Improved tolerance to high temperature conditions, termed acclimatisation, can occur in about two weeks of exposure and this is aided if heavy work is undertaken. Acclimatisation results from an increase in the body's ability to sweat, so enhancing evaporative cooling [2.6]. However, the regular intake of water and salt is still of great importance.

Heat loss by evaporation can be artificially enhanced by wetting the body. This is, for instance, an important consideration in establishing the thermal conditions for a swimming pool, while in hot dry climates it can contribute usefully to cooling processes. However, it can also be responsible for the dangers of 'exposure' in wet conditions, e.g. on mountains or sailing, particularly if the waterproofing of clothing is inadequate.

The relative rates of heat loss from the body by convection, radiation and evaporation will depend on the external temperature conditions, as summarised in Fig. 2.2.

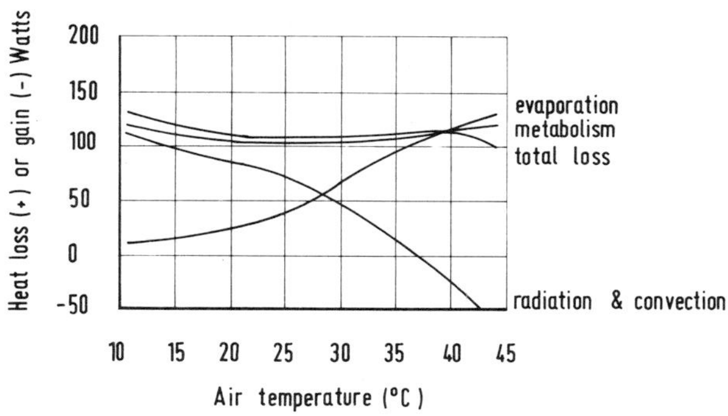

Fig. 2.2. Relative rates of heat loss from the body with increasing air temperature (adapted from Billington [2.7]).

There are other voluntary modes of control. These include the posture of the body; curling up reduces heat loss by reducing the surface area, while posture which creates the maximum surface area and free air movement around the body is appropriate in hot conditions. Clothing is an additional artificial means of affecting heat losses and gains to the body (see Section 2.2.4).

2.2.3 Thermal Parameters

The thermal parameters which will affect heat transfer to or from the body are

(a) air temperature,
(b) mean radiant temperature,
(c) air velocity and
(d) relative humidity.

It is these four parameters which must be controlled to appropriate values if desirable thermal conditions are to be achieved.

(*a*) *Air temperature* will affect the heat loss through natural convection from the body. If the air temperature is above the skin temperature, as in front of a warm air convector heater, then there will be heat gain to the body.

(*b*) *Mean radiant temperature*, i.e. the averaging of all the surface temperatures (walls, floors, etc.) around the body, will affect the radiant loss or gain to the surface of the body. Standing by large areas of cold glazing will produce high radiant losses to those surfaces, while a coal fire will produce high gains to the side of the body facing it. In typical conditions in well insulated buildings the mean radiant temperature is usually similar to the air temperature, and is not normally sensed directly. However, when asymmetric radiation conditions occur or there are extremes of surface temperatures then radiation effects become noticeable.

(*c*) *Air velocity* (in conjunction with the air temperature) will affect the heat lost or gained from the body by forced convection. The air velocity also affects the layer of air trapped close to the body surface. At very low air velocities (i.e. less than about 0·1 m/s) a stagnant layer of air builds up around the body which is not removed by air movement, and discomfort notionally described as 'stuffiness' or 'oppressiveness' can build up. The air movement values for 'freshness' are usually considered to lie between about 0·15 and 0·3 m/s [2.8]. Above 0·4 m/s, for typical internal air temperatures, discomfort can occur because the heat loss from the body rises, particularly at the extremities such as the hands and feet, or areas of exposed skin such as the neck. Localised low air movement, if associated with low air temperatures (i.e. draughts), is a common form of thermal discomfort. High air movements, usually encountered only outside buildings but also present near poorly designed ventilation outlets or doorways, can lead to large heat losses from the body as a whole. In external conditions around 0 °C, an increase in wind velocity from 1 to 2 m/s is equivalent to a drop in air temperature of about 3 °C [2.9]. Hence under such conditions air velocity becomes an important factor in human survival, and appropriate protective measures in the form of wind shielding around the body are vital. However, for warm temperature conditions or high activity air movement is desirable because it increases heat loss.

(*d*) *Relative humidity* (RH) affects the rate at which moisture can evaporate from the skin. If the RH values are kept within a certain range then this evaporation process can take place readily at rates governed by air temperatures and perspiration rates. This range is generally considered to lie between 40 % and 70 % [2.8]. Air movement is again important because in still conditions the layer of air close to the skin rises in moisture content and the evaporation process is hindered. This partly accounts for the 'stuffiness' effects previously referred to.

For RH values higher than 70 %, discomfort can begin to occur, especially in higher temperature conditions when heat loss is very dependent on evaporation. In buildings, mould growth and condensation effects can also become problematic at these levels. Also, the insulation value of clothing decreases at high RH values.

At low RH values (i.e. less than 40 %, but particularly below 30 %) other types of discomfort can occur (e.g. [2.10]). Drying out of the nasal passages leads to a more ready ingress of spore and dust particles into the respiratory system and an increase in throat and chest ailments, while sinuses can be affected causing headaches. Also a drying out of the skin will occur, leading to discomfort and possible aggravation of skin disorders, particularly affecting sensitive areas such as the eyes. These effects are all non-thermal, but are relevant as RH must be controlled for thermal comfort.

The four thermal parameters all interrelate. Comfort may be experienced for a certain set of values but if, for instance, the air temperature is raised, then the radiant temperature should be reduced or the air velocity raised to increase losses to compensate. Relative humidity and air temperature interrelate; overheating the air can lead to low relative humidities and insufficient heating to high relative humidities (see Section 1.7.2). Also, at different temperature conditions the parameters have differing significances. Air movement increases in importance at low air temperatures (e.g. complaints about draughts usually occur in winter) while relative humidity has most significance at high temperatures as it affects the major heat loss mechanism. The way in which these parameters interrelate becomes clearer when thermal comfort indices are described (Section 2.3).

2.2.4 Clothing

The skin and fatty subcutaneous tissue have an inherent thermal insulation value which acts as a natural buffer to the loss of heat from the bloodstream to the skin surface. The surface layer of hairs can affect this value, increasing insulation when raised.

As human beings are inherently tropical animals, their natural insulation layers are inadequate for many of the regions they now inhabit (in comparison, for instance, with polar bears or seals), and hence there is a requirement to increase this insulation using clothing. The fabric of clothing traps layers or pockets of air close to the skin, either in the clothing itself or between the fabric and the skin. The insulation value of clothing is measured in terms of thermal resistance, in units called togs, where 1 tog has a value of 0·1 m^2 °C/W. A more convenient unit is the clo (where 1 clo is 0·155 m^2 °C/W) because it gives convenient numerical values, e.g. 1 for a light suit. Typical values for a range of clothing are given in Table 2.2.

From the table it is clear that the heat loss rates from the body can be readily altered by appropriate clothing. Internal temperatures of buildings could be lowered if inhabitants compensated for this by increasing the clo-value of their attire. However, there are limitations to this action as the hands and face must usually be exposed and these will suffer discomfort with restriction in dexterity even if the rest of the body is heavily clad.

Table 2.2 Typical clo-values (after Fanger [2.4])

Attire	Clo-value
Naked	0
Shorts	0·1
Light summer clothing	0·5
Indoor clothing	1·0
Heavy suit	1·5
Suit plus overcoat	2
Polar clothing	3–4

The clo-value is considerably reduced if air is forced through or moisture enters the clothing. Hence, external clothing for hostile weather conditions should avoid wind and rain penetration. This must be achieved, however, whilst still allowing water vapour from perspiration to escape otherwise clothing becomes damp; this has led to the development of specialised fabrics [2.11].

2.2.5 Variation of Temperature over the Body

The body core temperature is maintained at approximately 37 °C. However, there is considerable variation in temperature throughout the body, in particular at the extremities (fingers and toes) as cooling of the bloodstream occurs before it reaches these regions where there is also little natural insulation present. This variation can be described by measuring skin temperatures. Typically, for an air temperature of 21 °C the forehead may be at 34 °C while the hands are at 29 °C and the feet at 26 °C [2.12].

Hence, if localised cooling of the body takes place, for instance by air movement, then regions such as hands and feet are susceptible to sensations of discomfort because their temperatures can drop more readily, while the insulation value of the clothing is often less, or does not exist, in these regions.

Complaints of cold feet and ankles are commonplace and should be treated seriously as general comfort cannot be achieved with this localised discomfort. The effect is usually enhanced by vertical temperature gradients in a space, and cold air moving across floors through downdraughts from windows or air movement under doors (e.g. [2.12]). The temperature of the hands is particularly important where manual work is to be performed, for the feeling and sensation in the fingers becomes dulled as the temperature drops. This not only has manipulative and resulting efficiency implications, but can cause safety hazards. Extremes of working environments, for example on oil pipelines in subzero conditions, present very distinct problems.

Skin surface temperatures can rise by localised overheating of the body. This often occurs through radiant sources affecting the face and head, e.g. radiant heating from some work processes such as melting and casting metals or from the sun. Recommendations of maximum radiation levels are available [2.8], while for extreme conditions localised protection, or controlling of exposure, should be introduced.

2.3 SCALES OF COMFORT

The body activity level and clothing combined with the four thermal parameters—air temperature, mean radiant temperature, air velocity and relative humidity—interrelate in the creation of conditions which may be considered by human beings as being thermally comfortable. Much work has been undertaken to establish the nature and quantification of the interrelations and to produce a single figure index to act as a summation of the factors. The first major index of this kind, used particularly in the USA and Canada, is the effective temperature and its development, the corrected effective temperature.

2.3.1 Effective and Corrected Effective Temperature Scales

The original effective temperature scale devised by Houghten and Yaglou in 1923 [2.13] combined the three factors of air temperature (i.e. dry bulb temperature), relative humidity in the form of the wet bulb temperature (see Section 1.7.2) and air velocity. Subjects were exposed to a certain set of conditions generated artificially and were asked to assess subjectively their sensations of comfort in terms of warmth and cold. The parameters were collated in graphical form so that conditions producing equal sensations of comfort could be assigned equal effective temperature values on a chart. Two such charts were generated; the first for persons stripped to the waist, known as the basic scale; the second for persons lightly clad, namely the normal scale. Both scales are for sedentary activity.

The original effective temperature scale did not consider radiation effects. Vernon and Warner in 1932 [2.14] modified the scale to take account of radiant conditions by substituting for the dry bulb temperature the globe thermometer temperature as read from a thermometer at the centre of a 6 in blackened copper sphere, and called the new scale the corrected effective temperature (CET) which is shown in Fig. 2.3.

The CET chart has two vertical axes, one for the globe or dry bulb temperature and the other for the wet bulb temperature, with curved lines running across representing values of air velocity. If the globe and air temperature measurements are similar then either value may be used; if not then the globe reading is used. If certain globe and wet bulb values are recorded, e.g. 25 °C and 15 °C for the normal scale, then a line drawn between these two will intercept the velocity curves. For a given velocity value an effective temperature value can then be read off. For this example, at 0·1 m/s the effective temperature is 21 °C while at 2·0 m/s the value has dropped to 18·5 °C.

The overall effects demonstrated by the effective temperature charts can be seen by fixing certain of the parameters. If the air temperature is at, say, 25 °C a lowering of the wet bulb temperature has the effect of lowering the effective temperature, as in physical terms more evaporation of perspiration can occur and cooling is enhanced. This effect is less important at lower air temperatures, hence the slight flattening of the effective temperature curves in this region.

For fixed globe and wet bulb temperatures it is seen that an increase in air velocity lowers the effective temperature, as heat losses rise through forced convection. This effect is more pronounced for lower temperatures, hence a spreading out of the velocity lines, whilst at higher temperatures the effect becomes reversed (i.e. a crossover of the velocity lines occurs) because of heating of the body surface by the air. On the basic scale (i.e. for minimal clothing) the effects of changing conditions are more pronounced as the buffering effect of clothing is not present.

The effective and corrected effective temperature scales are now considered to over-estimate the effects of relative humidity under typical

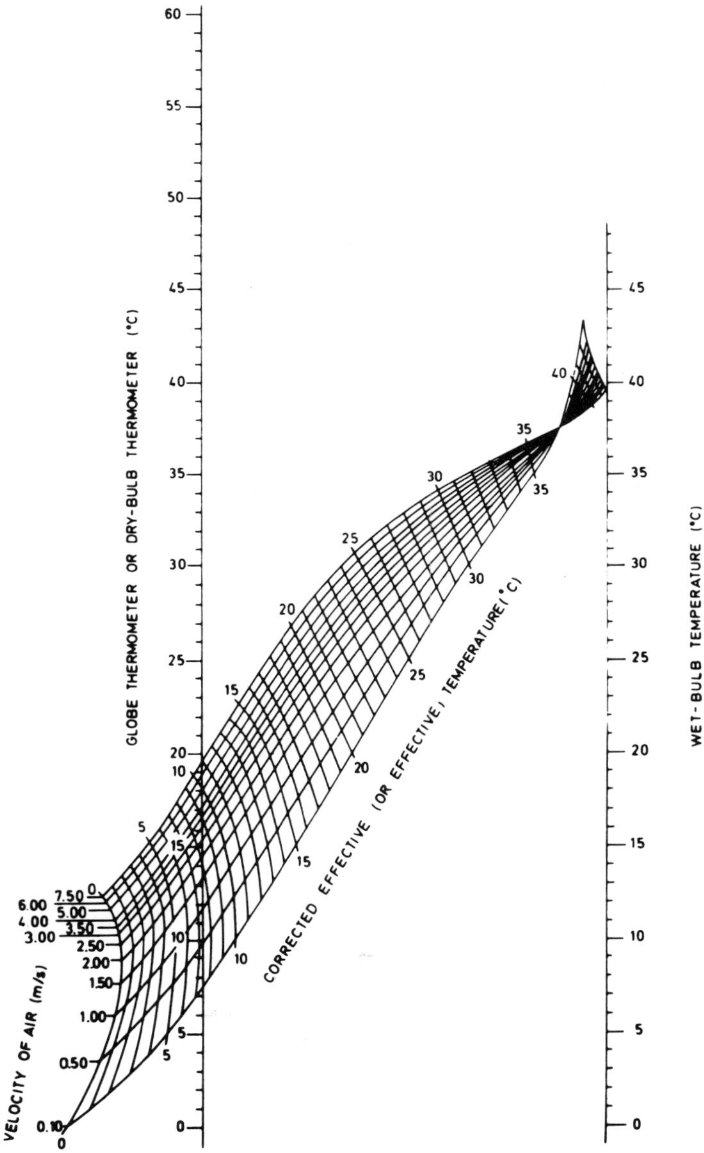

Fig. 2.3. The normal scale of corrected effective temperature (Ellis *et al.* [2.15], reproduced by permission of the *British Journal of Industrial Medicine*).

comfort conditions and to have short-comings because they apply to people only in sedentary activity [2.16].

2.3.2 Fanger Comfort Graphs

P. O. Fanger in studies mainly at the Technical University of Denmark [2.4] assessed human comfort through a thorough analysis of the human heat balance equation (Section 2.2.1). From the heat balance or 'comfort equation' the conditions were calculated under which heat gain and loss were equal for given variations of the parameters of activity, clothing, air and radiant temperatures, relative humidity and air movement. Balance can be achieved with extremes of parameter values, but this would not represent 'comfort' so some restrictions were put on certain of the parameters involved, e.g. the skin temperature, to create realistic 'comfort' conditions. The information obtained is represented through two sets of graphs.

The first of these assumes that the air and mean radiant temperatures are equal. The graphs show lines of equal 'comfort' for given values of wet bulb, air temperature and air velocity (relative humidity lines derived from the dry and wet bulb temperatures are also traced on the graphs for convenience). Each graph is for a given activity (sedentary, medium or high) and a given clothing value (nude, light, medium or heavy), creating twelve individual graphs in all. A set of these for varying activities with medium clothing is shown in Fig. 2.4.

For instance, for medium activity and medium clothing an air and radiant temperature of 15 °C, combined with 40 % RH, is predicted to give comfort at an air velocity of 0·1 m/s. However, if the air velocity is increased to 0·5 m/s (RH remaining at 40 %) the air and radiant temperatures must be increased to 18 °C for comfort to be maintained.

These graphs in general show that relative humidity has a fairly small effect on the sensation of comfort. A change of RH value from 0 to 100 % can be compensated for by temperature decreases of 1·5–3 °C. It must be stressed that this applies to *comfort* conditions and that as soon as discomfort on the warm side is experienced, relative humidity is of greater concern as demonstrated by its importance in heat stress indices (Section 2.3.5). As has already been stated (Section 2.2.3) there are other aspects, apart from thermal comfort, that make the control of RH value to the range 40–70 % important.

The second approach by Fanger to summarising comfort conditions takes into account differences in air temperature and mean radiant temperature. A set of twelve graphs is again produced for different

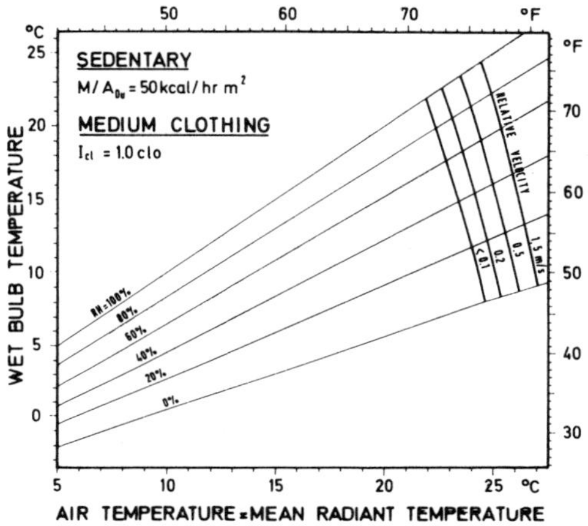

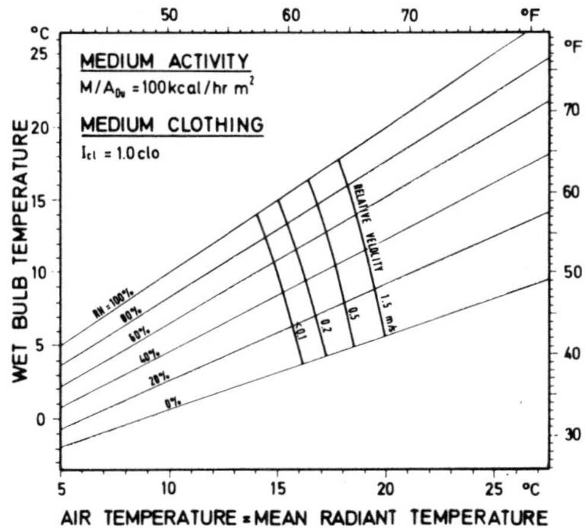

Fig. 2.4. Fanger comfort graphs of air temperature (equal to mean radiant temperature) versus wet bulb temperature with relative air velocity as parameter for persons wearing medium clothing (clo-value = 1·0) for sedentary, medium and high activities (metabolic rates ≃ 105 W, 210 W and 315 W, respectively) (from *Thermal Comfort* by Fanger [2.4], © Fanger 1970, reproduced by permission of the McGraw-Hill Book Co.).

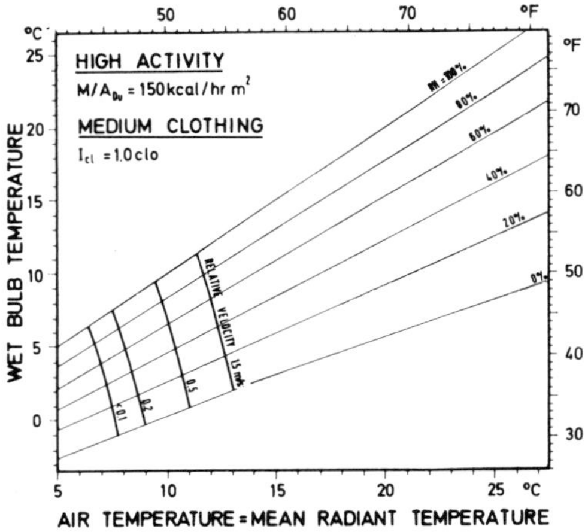

Fig. 2.4—contd.

conditions of clothing and activity, but now mean radiant temperature is plotted against air temperature for different velocity values, with relative humidity kept constant at 50%. The graphs for various activities and medium clothing are shown in Fig. 2.5.

From these graphs one can read values of air temperature, mean radiant temperature and air velocity predicted to create comfort conditions. For instance from the sedentary activity and medium clothing graph, at 25 °C air temperature a vertical intercept to the 0·1 m/s velocity curve gives a mean radiant temperature value of 20 °C. So these three parameters should be providing comfort conditions for these values of clothing and activity. If the air velocity rises to, say, 0·5 m/s the mean radiant temperature would need to be 23·5 °C for comfort.

The shift in the velocity lines to lower temperature values as activity increases is to be expected; a similar effect occurs as the clothing value increases. The cross-over of velocity lines for high air temperatures is the result of heating by forced convection, as occurred on the corrected effective temperature chart. This effect does not occur for high mean radiant temperatures as the radiant heat transfer process is not affected by air movement.

In typical internal environments, air velocity values are 0·1–0·2 m/s and mean radiant temperature values usually differ by about 1–2 °C from the air temperature. Hence comfort values of temperatures can be extracted from the graphs. However, if an extreme is being dealt with, e.g. radiant heating

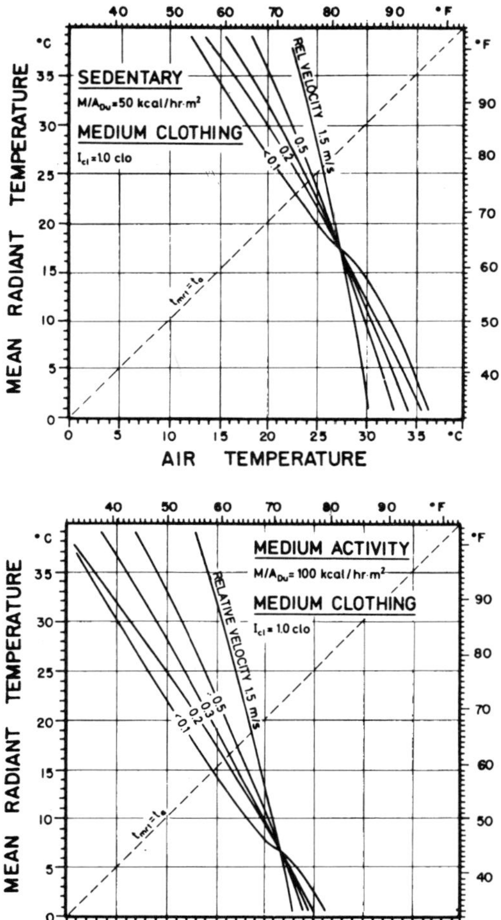

Fig. 2.5. Fanger comfort graphs of air temperature versus mean radiant temperature for relative humidity of 50 % with relative air velocity as parameter for persons wearing medium clothing (clo-value = 1·0) for sedentary, medium and high activities (metabolic rates ≃ 105 W, 210 W and 315 W, respectively) (from *Thermal Comfort* by Fanger [2.4], © Fanger 1970, reproduced by permission of the McGraw-Hill Book Co.).

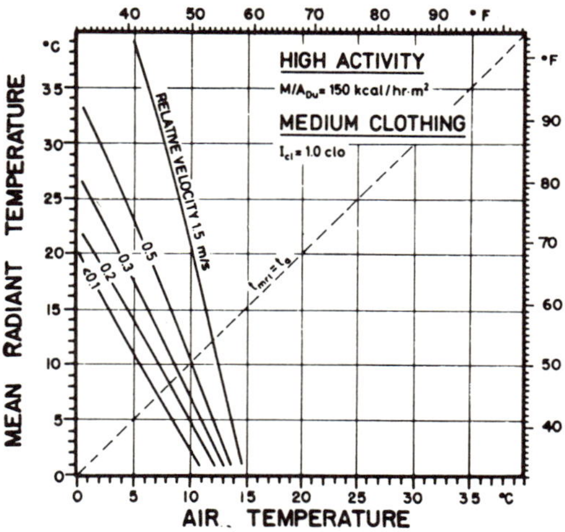

Fig. 2.5—contd.

or large areas of cold windows, then the appropriate balance of mean radiant temperature and air temperature can be determined. The graphs are particularly useful for determining conditions for buildings with special activities and clothing, e.g. swimming pools or certain industrial environments.

Radiant conditions are often asymmetric and a single figure value read from the comfort graphs can be misleading when applied to real conditions. In such cases the radiant fields require a more thorough analysis [2.4].

It should be noted that the Fanger graphs are aids for predicting comfort conditions whereas the effective temperature plots are giving conditions of equal sensation, whether they be comfortable or not, and that the effective temperature plots on their own do not predict conditions which are regarded as comfortable.

2.3.3 Resultant Temperature

The resultant temperature is the index of thermal comfort now recommended for use by the Chartered Institution of Building Services (CIBS) in the UK [2.8]. The index is the temperature of a globe thermometer (i.e. black bulb thermometer) of 100 mm (4 in) diameter. This combines air temperature and mean radiant temperature for 'still' air conditions (i.e. 0·1 m/s) in the proportions given approximately by:

$$\text{Resultant temperature} = \tfrac{1}{2}(\text{Air temperature}) + \tfrac{1}{2}(\text{Mean radiant temperature})$$

This is considered to be a response to the combination of these two parameters similar to that of the human body [2.17]. For air movement above 0·1 m/s adjustments should be made to the resultant temperature (Fig. 2.6). For example, an air speed of 0·2 m/s is compensated for by a 1 °C increase in the resultant temperature. When quoting resultant temperature values the relative humidity is assumed to lie in the comfort range, i.e. 40–70 % (see Section 2.2.3).

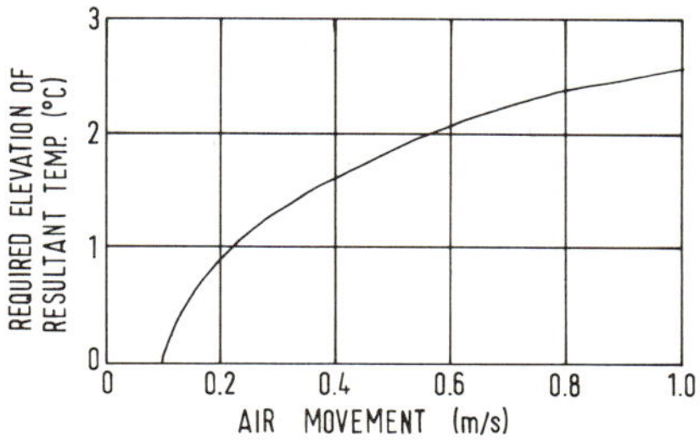

Fig. 2.6. Corrections to the dry resultant temperature to take account of air movement (from CIBS Guide A1 [2.8], reproduced by permission of the CIBS).

Recommended temperature values quoted for internal environments and comfort conditions in the CIBS Guides are resultant temperature values rather than any other comfort index. When air and mean radiant temperatures are similar then the resultant temperature approximates to the air temperature, while for most conditions its value is close to that of the environmental temperature, i.e. the parameter used in heat transfer calculations (see Section 1.6.2), and the two may therefore be interchanged. The resultant temperature is a value generally more readily measured than predicted. During design the required environmental conditions for the new building should be specified, but the fine tuning of these conditions is often finally achieved by measurement and adjustment of controls in the completed building. The design process should take into account pronounced differences between air and radiant temperatures in order to minimise factors which create difficulties in control, e.g. direct solar gain or excessive air movement, so that the specified resultant temperature is in practice readily achievable.

2.3.4 Standard Effective Temperature

The thermal index adopted by the American Society of Heating, Refrigeration and Air Conditioning Engineers (ASHRAE) [2.16] is the standard effective temperature, which was developed by Gagge and his colleagues [2.18]. The derivation and definition of this index is relatively complex and the original work or ASHRAE documentation should be referred to for further details.

The standard predicted comfort conditions are for occupants wearing clothing of 0·4 to 0·6 clo-value and in sedentary activity. The predictions are similar to those arising from the work of Fanger (Section 2.3.2).

2.3.5 Other Scales and Indices

There are a large number of other indices which have been developed and used. These are now mainly of historical and theoretical interest. They include the equivalent temperature [2.19], equivalent warmth [2.20] and the equatorial comfort index [2.21].

There are other indices which are generated for the assessment of thermal extremes. For hot conditions these include the wet bulb globe thermometer (WBGT) index [2.22], which is now adopted by the American Conference of Governmental Industrial Hygienists (ACGIH) for threshold limit values [2.23]; the Belding and Hatch heat stress index [2.24]; and the predicted 4-hour sweat rate (P4SR) [2.25]. All these indices put major emphasis on the relative humidity value or wet bulb temperature because of the importance of evaporation as a heat loss mechanism. For cold conditions the major index is the windchill factor which assesses the extra cooling effect that wind creates in cold conditions [2.9].

2.3.6 Recommended Conditions for Comfort

Much discussion has taken place over the meaning of the term 'thermal comfort' and of the more general interpretation of 'comfort' from all aspects (i.e. lighting, acoustic, general well-being, etc.) [2.4, 2.12, 2.16, 2.26]. From the point of view of the building designer it is necessary to be able to specify thermal conditions for building interiors which will satisfy the majority of occupants and to avoid those conditions which are likely to lead to discomfort and complaints. Most studies on comfort have been confined to people at rest or in sedentary occupations. Bedford [2.12] expressed comfort in terms of temperature values for which 70 % of persons tested did not experience discomfort. Recommendations were that an effective temperature range of 14–17 °C should provide comfort in winter conditions in the UK for people lightly dressed involved in light activity. For completely sedentary activity the figures are 1–2 °C higher. In summer conditions the values were 16 °C to 22 °C. These values are 2 °C to 3 °C lower than would be expected in North America or Canada, and current recommendations would suggest that they are rather low for the UK [2.8].

From the Fanger comfort graphs (Section 2.3.2) in low air movement conditions (0·1 m/s) with air and mean radiant temperatures equal, comfort would be experienced for sedentary persons in light clothing at an air temperature of 26 °C, while if the activity were increased to medium the value would be 20 °C, or if the clothing increased to medium, 23 °C. Similar values are obtained using the standard effective temperature.

Current recommendations in the UK (CIBS [2.8]) suggest that a resultant temperature lying between 19 °C and 23 °C should provide comfort conditions for the majority of people in sedentary or light activities, assuming that fine tuning of comfort invariably takes place through altering attire and activity. These values are slightly lower than those predicted by Fanger.

The resultant temperature is made up from radiant and air temperature components and a large imbalance of the two could still provide a resultant temperature as recommended but discomfort would be likely. Consequently, it is also suggested that the mean radiant temperature does not exceed the air temperature by more than 8 °C or fall below it by more than 5 °C. Optimum comfort is considered to occur when the mean radiant temperature is held a few degrees higher than the air temperature, generating a 'fresh' rather than 'stuffy' environment, although there is little evidence for this view.

Comfort in active occupations is less readily specified. Local and personal control is often undertaken (removing clothing, enhancing air movement, etc.) while appropriate conditions can be derived, using for instance the Fanger comfort graphs. Health and safety legislation may also specify minimum or maximum temperature values.

In buildings where particular activities take place (e.g. swimming pools and sports halls) the thermal conditions have to be suited to certain conditions of clothing and metabolic rate and suitable conditions can again be predicted from the Fanger graphs. Invariably, however, an interested group or association connected with the activity involved produces guidelines for recommended conditions, although these may be of minimum information, e.g. a single air temperature value.

In conditions of high activity and high temperatures, often found in industry, recommendations or legislation apply which govern the length of time activity can take place. The WBGT index is now the most widely used index for such situations as recommended by the ACGIH [2.23].

Fanger [2.4] looked at the variations in comfort conditions experienced by people of different nationalities, age, sex and body build. He found that there were negligible differences in the conditions that people considered as thermally comfortable. He also noted that there were no variations with menstrual cycles, food intake or circadian rhythms.

However, experience seems to imply that there are differences in the conditions that people consider as comfortable; but these can usually be explained by variations in clothing or activity. For instance, female office workers are often reported to request warmer conditions than their male counterparts. This could be explained by the former group wearing clothing of lower clo-value. Older people tend to favour higher temperature conditions than younger, but this may be caused by differences in activity level.

2.3.7 Causes and Effects of Thermal Discomfort

Thermal discomfort is considered to occur when people are not satisfied with thermal conditions, but conditions are not severe enough to cause medical symptoms associated with extremes. It can be created by very minor effects, but still generate considerable annoyance.

The simplest cause of thermal discomfort arises from the thermal parameters being set at such values that their combination lies outside the range of comfort or that an individual parameter has an extreme value. Such conditions can usually be readily assessed by observation or measurement and alterations made to control mechanical systems or building fabric to shift conditions into the comfort region.

More commonly, however, discomfort arises from uneven or asymmetric conditions. For instance, air temperature differences between head and feet should preferably not exceed about 3 °C [2.8]. Thermal gradients due to convection (see Section 5.3.2) sometimes makes this difficult to achieve. Localised air movement, particularly if the air is cooler than the overall air temperature, should be avoided. This is often difficult to prevent in cold conditions when there are cracks around windows and doors and downdraughts from cold windows. Excessive localised heating should be avoided, e.g. warm air fed directly towards people or a close source of high temperature radiant heat. Direct heating from the sun through windows causes discomfort and radiant heating to the head in particular should be controlled [2.8]. Lack of air movement, particularly with high air temperatures and high relative humidities, leads to complaints of stuffiness.

These aspects are summarised in Fig. 2.7. Many of the causes of discomfort can be eliminated by appropriate attention to detail at the design stages of a building, internal space or mechanical installations.

Deviations from comfort can lead initially to general dissatisfaction with

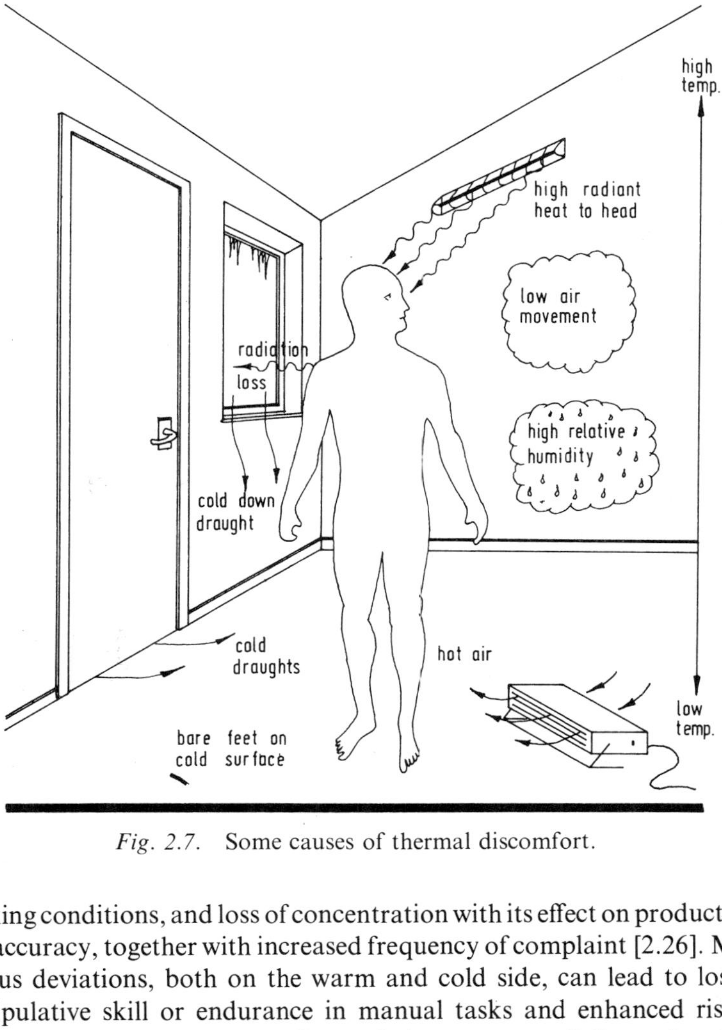

Fig. 2.7. Some causes of thermal discomfort.

working conditions, and loss of concentration with its effect on productivity and accuracy, together with increased frequency of complaint [2.26]. More serious deviations, both on the warm and cold side, can lead to loss of manipulative skill or endurance in manual tasks and enhanced risk of accidents [2.12, 2.26]. Some effects of the extremes of discomfort have already been mentioned (Section 2.2.2); these are further discussed elsewhere [2.6, 2.27].

2.4 AIR QUALITY

Ventilation is required to introduce clean air into an environment and to remove air contaminated by occupants and their activities. The main contaminants from occupants are carbon dioxide, water vapour and odours, while common activities generate tobacco smoke, water vapour and combustion products from appliances. Hence, the primary function of ventilation is one of general comfort and health through the generation of suitable conditions for respiration. However, thermal comfort is closely linked to ventilation through aspects of air movement around people, through the removal of excess heat and through the provision of fresh air at an appropriate temperature.

Ventilation rates are usually quoted as either a volume of air to be introduced per second (e.g. litres/s) or else the number of times a volume of air equivalent to the size of the room or building is introduced per hour and expressed as 'room volumes per hour'. The room volume rate specification does not mean that all the air in a room is totally replaced a given number of times per hour. Rather it means that the total volume of air moving through the space in an hour is equal to the room volume multiplied by the room volumes per hour value. The removal of contaminants is therefore not a total purging as would occur if all the air were removed (creating a vacuum) followed by it being replaced, but rather a dilution process whereby a mixing of contaminated air with fresh air occurs and over a period of time the contaminant is 'diluted'. The actual dilution rate depends on the ventilation rate, the production rate of the contaminant and the mixing that takes place. Descriptions of the means by which ventilation is achieved occur in Chapter 4.

2.4.1 Natural Air Contaminants from People

The composition of dry air is given in Table 2.3. Alterations to this composition by human activity are to reduce the oxygen content, to increase the carbon dioxide content and to add water vapour and odours. A small decrease in the oxygen concentration (e.g. 1–2 %) has little effect on human beings, whereas an increase in carbon dioxide levels is manifest in discomfort such as headaches and lassitude. Hence the removal of carbon dioxide is of particular importance.

Table 2.3 Composition of dry air

Constituent	Percentage by volume
Nitrogen	78·08
Oxygen	20·94
Carbon dioxide	0·03
Other gases	0·95

An individual produces carbon dioxide at a rate ranging from 0·004 litres/s at low activity to 0·03 litres/s for high activity. The atmospheric concentration of carbon dioxide can reach 4% for short periods without undue problems, but 0·5% is usually the accepted limit. The air supply rate in a typical occupied room to achieve this is 1 litres/s [2.28].

Water vapour is given off at 0·03–0·04 kg/h from an active person. The main significance of this is the build up of relative humidity values which can lead to problems of mould growth and condensation. These problems are much exacerbated by activities creating water vapour such as washing and cooking (Section 2.4.2).

Body odours are difficult to identify and quantify, but represent the most important natural contaminant governing ventilation rate requirements. The generation and endurance of these odours are dependent on a number of factors including temperature, nature of activities and standards of cleanliness. The acceptability of levels is also dependent on adaptation to concentrations; the olfactory senses rapidly lose their sensitivity when exposed to a constant or slowly changing level of odour. Hence levels of acceptability of odours vary depending on whether the person is continuously exposed or newly exposed to the contaminated space.

The air supply rate required for odour removal is typically five times greater than that required for carbon dioxide removal [2.29] and hence ventilation rates under normal conditions of occupation are governed by odour removal. Recommended rates are summarised in Fig. 2.8 for differing levels of acceptability to occupants and outsiders.

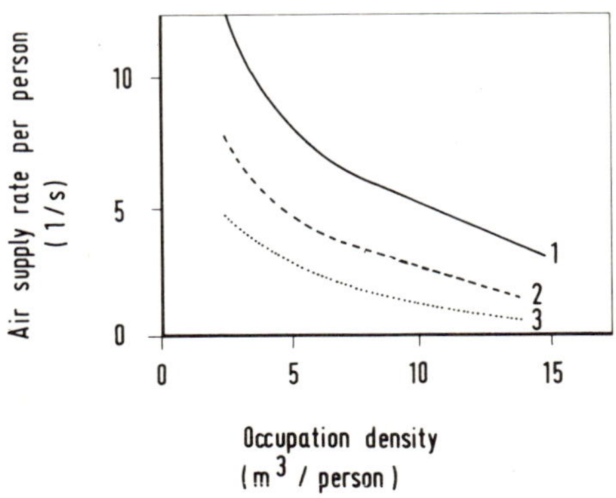

	Occupant	Outsider
1	Good	Acceptable
2	Intermediate	
3	Fair	Objectionable

Fig. 2.8. Air supply rates for odour removal for differing occupation densities (adapted from BRE Digest 206 [2.30]).

2.4.2 Additional Contaminants from Activities

Tobacco smoke is the contaminant most widely encountered in occupied spaces. The levels of acceptability from a subjective comfort viewpoint vary considerably, especially between classes of 'smokers' and 'non-smokers', and a blanket coverage for contamination removal cannot satisfy all parties. However, in general, the ventilation rates where smoking is permitted are usually raised by about one-third over non-smoking values, as shown in Table 2.4.

Table 2.4 Minimum ventilation rates [*2.28*]

Air space per person (m^3)	Fresh air supply per person (litres/s): recommended minima	
	Non-smoking	Smoking
3	17·0	22·6
6	10·7	14·2
9	7·8	10·4
12	6·0	8·0

Artificially created water vapour is particularly problematic in winter in dwellings, where activities such as washing and drying of clothes, bathing and showering take place. Table 2.5 summarises the water vapour emissions produced by such activities.

The main consequence of this is the resulting high levels of relative humidity which can be generated. These levels are temperature dependent, and so the standard of the building fabric and heating systems influence the problem. High relative humidity values can result in raised bacterial and spore levels (affecting respiratory ailments and, in the past, tuberculosis

Table 2.5 Water vapour levels given off by household activities for a family of five [*2.30*]

Household activities	Typical water vapour emission (kg/day)
Cooking	3
Bathing	1
Dishwashing	1
Clothes washing	0·5
Clothes drying	5

transmittance), mould growth on walls and contents and condensation (see also Section 1.7.2). The interrelation of condensation risk, activity, temperature and fabric have been studied theoretically [2.31], leading to optimum recommended ventilation rates for the alleviation of the problem. The significant information from these studies is that too high a ventilation rate will cause internal temperatures to fall and relative humidity values to rise (unless excessive heating is used), hence exacerbating the problem. The optimum rate is a balance between removing the water vapour contaminant by ventilation but not lowering the air temperature values unduly.

Very low relative humidity values have been mentioned previously (Section 2.2.3) as affecting human health and comfort. The mucous membranes of the respiratory system become dehydrated and dust and spore particles enter more readily, while materials themselves (e.g. carpet fabrics) become more susceptible to breakdown into fine dusts. Hence nose and throat irritations and respiratory ailments, in particular those asthmatic in nature, become more prevalent, as does dried skin with associated risk of allergy, irritation and infection, particularly to the eyes (e.g. [2.26]).

The build-up of static electricity and resulting electrical shocks can also occur at low relative humidities. In this case the problem is due to a deficiency, and one obvious cure is to add water vapour to the atmosphere. If the building services do not provide humidification then this can be achieved with local humidifiers, e.g. moist 'blotting paper' in a reservoir of water hung on a radiator or a fan-assisted evaporation humidifier, or to a certain extent with plants and flowers. Lowering temperatures is also a direct and simple approach; the dry air syndrome is usually associated with overheated dwellings or offices.

Boiler and heater combustion products such as carbon dioxide, carbon monoxide, sulphur dioxide and water vapour are the final class of contaminants commonly encountered. Combustion devices are usually designed to have inbuilt ventilation or come under regulations for permanent ventilation into the space, e.g. by air bricks or ducted air. The appropriate ventilation rate should therefore be provided. However, a number of problems can occur with combustion appliances. Portable units such as paraffin heaters and bottled gas heaters are often used without allowances made for extra ventilation; excessive condensation is the usual result. Air bricks for permanent ventilation may become blocked. Inadequately maintained or damaged appliances and damaged flues can lead to dangerous circumstances arising, particularly if the toxic products are given out during the night when occupants are asleep. In the extreme cases death can result. In general, though, such circumstances are not under the control of the building designer.

2.4.3 *Ventilation Rates*

For convenience ventilation rates are usually expressed in room volumes per hour (RVH) of fresh air (as opposed to recirculated air) for particular rooms or building types. The fresh air supply rates previously quoted (Fig. 2.8 and Table 2.4) combined with information relating to other contaminants (e.g. water vapour or smoke) are brought together for rooms of particular activities and typical volumes to give these recommended RVH values. Examples are given in Table 2.6. However, if uncertainty

Table 2.6 Typical ventilation rates (adapted from CIBS Guide B2 [2.32])

Space	Recommended fresh air ventilation rate in room volumes per hour (RVH)
Boiler houses	15–30
Bathrooms	6
Cinemas	6–10
Dining halls	10
Kitchens	20–60
Laboratories	4–6
Living rooms	1–4
Offices	4–6
Theatres	6–10
Toilets	2–6

exists over a recommended ventilation rate, due to perhaps unusually large or small volumes or densities of occupation, then the basic criteria of air supply rate per person in a known volume should be used to determine the ventilation rate. It should be noted that new buildings need higher ventilation and heating rates than older ones. This is to assist in the initial drying out from 'wet' construction processes and in the removal of possible noxious vapours arising from new materials (see Section 2.4.7).

2.4.4 *Threshold Limit Values*

There are a large number of airborne contaminants with possible toxic effects which can occur in industrial environments, and occasionally in other habitable environments. Many of these contaminants have designated to them a threshold limit value (TLV). This value is, in simple terms, a time weighted average concentration of contaminant which should not affect the health of the majority of people when exposed for a 40 h working week and a full working life. The philosophy behind the use of TLVs and detailed listings are to be found in the UK Health and Safety Executive Notes [2.33], the American Conference of Governmental Industrial Hygienists (ACGIH) Listings [2.23] or other individual government documentation. It should be emphasised that the TLV is not a dividing line between safe and unsafe working conditions, and that the best practice is to maintain atmospheric concentrations as far below the TLV as is reasonably practicable.

Contaminants are best controlled at source, but if they are ejected into the air then they must be diluted or removed either by general ventilation or by local extract ventilation (see Section 4.2). The rate of dilution or extraction is a combined function of the emission rate of the contaminant and its threshold limit value and is further discussed elsewhere (e.g. [2.16, 2.32]).

2.4.5 Air Ionisation

Atoms may become electrically charged due to the loss or gain of electrons and as such are termed positive or negative ions. These are created in the atmosphere by phenomena such as intense electrical fields (e.g. in thunderstorms), natural radioactivity, cosmic rays and the shearing of water droplets as occurs in waterfalls or fountains. These ions usually become attached to airborne dust or liquid droplets resulting in 'fresh' air with a natural ion content.

The biological effects of small negatively or positively charged particles in respired air have been a matter of some debate for many years. Dessauer [2.34] and associates suggested that negative ions produced a feeling of well-being and improved health conditions, whilst positive ions produced a rise in metabolic rate and blood pressure and caused symptoms such as headaches, nausea and general malaise. However, in extensive studies Yaglou *et al.* [2.35] found that they could not justify the use of artifical ionisation in ventilation systems.

There has been considerable conflicting evidence ever since these early studies. The American Society of Heating, Refrigerating and Air Conditioning Engineers commissioned a further study in 1962 [2.36] which confirmed the earlier work of Yaglou *et al.* [2.35], whereas Sulman [2.37], Kreuger and Reed [2.38] and Christian Bach [2.39] have confirmed positive findings. Interest has recently been further aroused by studies by Jones *et al.* [2.40] and by Hawkins and Barker [2.41].

It appears that some of the reasons for conflicting evidence arise from the types of subjects selected and the experimental controls imposed. Negative ions would seem to have little effect on persons who are normally fit and healthy, but are more likely to affect frail, less healthy and 'weather sensitive' persons. Also static field effects from clothing or surroundings can negate any effects which might have occurred.

There is certainly growing opinion to suggest that many totally controlled internal environments leave much to be desired in terms of comfort and health (e.g. [2.42, 2.43]). Much of this can be caused by inappropriate thermal, air movement, humidity or ventilation conditions, whilst lighting and acoustic effects can be contributory.

The possible effects of various concentrations of air ions are less tangible. The naturally occurring concentrations of air ions are modified when fresh air is confined in a room, particularly when there are numerous exposed 'plastic' surfaces (e.g. carpets, fabrics and finishes) which can hold static electrical charges. Furthermore, the naturally occurring concentration of air ions in fresh air is modified when the air is brought into buildings through metal or plastic ducts.

The detrimental effects of unnatural concentrations of air ions have yet to be established fully. It is probable that a marginal effect can never be isolated clearly experimentally to full scientific rigour because of many other factors which affect human behaviour and performance over extended periods of time.

2.4.6 External Atmospheric Contaminants

Ventilation specifications invariably use the term 'the introduction of fresh, clean air' into buildings. However, 'outdoor' air, particularly in the last century, has become not as 'fresh' or 'clean' as would be desirable. Major airborne pollutants have arisen through the burning of fuels and from large scale industrial processes producing in particular sulphur dioxide and soot. These gave rise to extensive 'smog' pollution in densely populated urban areas, e.g. Donora, Pennsylvania, USA in 1948 and in London in 1952 where excess deaths over the norm occurred [2.44, 2.45, 2.46]. Such events led to legislation such as Clean Air Acts and Environmental Airborne Standards, providing control and improvement of conditions.

More recent general environmental pollution problems have been generated by automobile exhaust products (e.g. [2.47]), in particular the resultant effects when combined with intense sunshine leading to photochemical smog. These problems initially arose in cities such as Los

Angeles and Tokyo, while in the 1970s large areas of Europe had extensive problems [2.48]. The major contaminant produced is ozone, and levels above the threshold limit value (Section 2.4.4) have been recorded for extensive periods of time [2.49]. Again, legislation by governments has been or is being introduced in an endeavour to curtail the problem.

Other common forms of external airborne pollutants affecting buildings are dust and dirt, produced particularly in urban areas from roadways and construction sites. Such airborne contaminants lead to problems for the building designer in the provision and control of ventilation. Natural ventilation transports such contaminants into buildings, and this may be undesirable, e.g. in offices or 'clean' work processes. Mechanical ventilation systems invariably remove particulate matter using various filtration systems, to the level required for the building under consideration (e.g. [2.10]), but poor maintenance can lead to problems. The removal of gaseous contaminants is more difficult and costly but can be achieved, for instance, with activated carbon filters.

2.4.7 Problems Associated with Low Ventilation Rates

Ventilation can account for a considerable proportion of the total energy running costs of a building (see Section 3.3) and it is therefore not unexpected that a reduction of ventilation rates is often recommended and put into practice as a means of reducing running costs. This may occur in a controlled fashion in a fully mechanically ventilated building, or be achieved to an unknown extent, for instance when windows and doors are 'draught proofed'.

As most general ventilation requirements are based on odour removal (Section 2.4.1) this may be the first aspect affected by reduced ventilation rates assuming that the original rate was similar to the recommended value. However, increased odour levels can be tolerated as occupants become insensitive to the new levels [2.50]. Techniques can also be employed for deodorising the air either by destruction of the airborne chemicals involved, e.g. by irradiation with ultra-violet light, or by masking with a more pleasant odour [2.50].

A potentially much more serious effect is the increase in relative humidity levels, particularly in densely occupied dwellings, and resultant mould growth and condensation which can damage building fabric and contents [2.31]. Also a build-up of carbon dioxide to high levels for long periods is potentially damaging to human health (e.g. [2.51]).

Other commonly produced contaminants can also rise to undesirable levels. High temperature appliances for heating or cooking can generate nitrogen dioxide which is a relatively toxic gas [2.51]. Tobacco combustion products may not be removed effectively thus exposing non-smokers to a greater risk to their health. An increase in general airborne bacteria and spore contaminants enhances the risk of cross-infection [2.16, 2.26].

There is also the possibility of the build-up of certain low level contaminants if ventilation rates are considerably reduced. Many building materials such as stone and brick exhibit low levels of radioactivity; the levels vary considerably with the source of the material. From such materials, airborne radioactive radon gas and associated products can be formed. Surveys of such contaminants in dwellings have been undertaken in both Scandinavia and the USA (e.g. [2.52]).

Certain plastics and resins give off low levels of organic vapours. For normal ventilation rates these are usually at very low concentrations; however, build-up can occur. Formaldehyde vapour, from urea formaldehyde (cavity foam insulation) and certain resins (e.g. in chipboard), has been monitored in dwellings in Denmark [2.53]. It can be responsible for eye, nose and throat irritations and allergic reactions [2.51], while carcinogenic effects from organic vapours involved cannot be ruled out [2.51].

Other building materials have generated concern over health. Asbestos is perhaps the most well known and its application in building construction has hence been considerably reduced. However, much asbestos is present in existing buildings and studies have been undertaken to examine the exposure risks to occupants (e.g. [2.54, 2.55]). These levels would naturally be increased with reduced ventilation. Health risks of building and construction materials in general have been reviewed by Kinnersly [2.56].

Finally, in many industrial environments contaminants are removed by general overall ventilation (as opposed to localised removal) by, for instance, the fitting of extractor fans in roofs. General indoor levels of contaminants could be increased with reduced ventilation, with possible health repercussions depending on the particular materials involved.

Hence it is shown that there are areas of concern associated with reduced ventilation rates. The topic of indoor air quality is receiving more attention, as illustrated by the World Health Organisation First International Indoor Climate Symposium held in Copenhagen in 1978 [2.43].

2.4.8 Thermal Conditioning of Ventilated Air

The relationship of air temperature and air movement to thermal comfort has been described (Section 2.2.3). Some care has to be taken in ventilation design to ensure that the general specified comfort conditions are not lost at points where ventilation air is introduced or removed.

Particular problems occur in naturally ventilated buildings in cold conditions where cold air infiltration around doors or windows or downdraughts from cold window surfaces cause undesirable draughts to occupants. Alleviation is assisted by 'draught stripping' although care must be taken to avoid reducing ventilation rates to such low values that contamination builds up excessively. The positioning of heating units can be arranged to counteract the effects of cold fresh air or downdraughts, for instance by siting radiators under windows.

Mechanical ventilation systems should introduce air at the appropriate temperature, in the correct position and at a suitable velocity so as not to disturb or create discomfort to occupants, particularly those sited close to the input point. The system should ventilate (and heat or cool) the whole space effectively. In hot conditions cooling by air movement can be achieved by increasing overall levels of ventilation, especially in naturally ventilated buildings. Alternatively, enhanced air movement is created by localised fans.

The production and control of ventilation by natural and mechanical means is discussed in Chapter 4 and the influences of heating and cooling systems in Chapter 5.

REFERENCES

2.1 BURBERRY, P.: *Building for Energy Conservation.* Architectural Press, London, 1978.

2.2 BANHAM, R.: *The Architect and the Well-tempered Environment.* Architectural Press, London, 1969.

2.3 NEWBURGH, L. H. (Ed.): *Physiology of Heat Regulation.* W. B. Saunders Co., New York, 1949.

2.4 FANGER, P. O.: *Thermal Comfort: Analysis and Applications in Environmental Engineering.* McGraw–Hill, New York, 1970.

2.5 BURBERRY, P.: *Environment and Services: Mitchell's Building Construction Series.* B. T. Batsford Ltd, London, 1975.

2.6 LEITHEAD, C. S. and LIND, A. R.: *Heat Stress and Heat Disorders.* Cassell & Co., London, 1964.

2.7 BILLINGTON, N. S.: *Building Physics: Heat.* Pergamon Press, Oxford, 1967.

2.8 CHARTERED INSTITUTION OF BUILDING SERVICES: *CIBS Guide A1: Environmental Criteria for Design.* CIBS Publications, London, 1978.

2.9 SIPLE, P. A. and PASSEL, C. F.: Measurements of dry atmospheric cooling in subfreezing temperatures: *Proceedings of the American Philosophical Society,* Vol. 89 (1945), p. 177.

2.10 CROOME-GALE, D. J. and ROBERTS, B. M.: *Air Conditioning and Ventilation of Buildings.* Pergamon Press, Oxford, 1975.

2.11 MORRIS, J. V.: Developments in cold weather clothing. *Annals Occup. Hyg.*, Vol. 17 (3–4) (1975), pp. 279–294.

2.12 Chrenko, F. A. (Ed.): *Bedford's Basic Principles of Ventilation and Heating*. H. K. Lewis, London, 1974.
2.13 Houghten, F. C. and Yaglou, C. P.: Determining lines of equal comfort. *Trans. Amer. Soc. Heat. Vent. Eng.*, Vol. 29 (1923), p. 163.
2.14 Vernon, H. M. and Warner, C. G.: The influence of the humidity of the air on capacity for work at high temperatures. *J. Hyg. Camb.*, Vol. 32 (1932), p. 431.
2.15 Ellis, F. P., Smith, F. E. and Walters, J. D.: Measurement of environmental warmth in S.I. units. *Brit. J. Ind. Med.*, Vol. 29 (1972), p. 361.
2.16 American Society of Heating, Refrigeration and Air-conditioning Engineers: *1977, Fundamentals Handbook*, Chapter 8. ASHRAE, New York, 1977.
2.17 Missenard, A.: Théorie simplifié du thermomètre résultant–thermostat résultant. *Chauffage et Ventilation*, Vol. 12 (1935), p. 347.
2.18 Gagge, A. P., Nishi, Y. and Gonzalez, R. R.: Standard effective temperature—a single index of temperature sensation and thermal discomfort. *Proceedings of the CIB Commission W45 (Human Requirements) Symposium*, Building Research Station, 13th–15th September 1972. HMSO, London, 1973.
2.19 Dufton, A. F.: *The Equivalent Temperature of a Room and its Measurement*. Building Research Tech. Paper no. 13, 1932.
2.20 Bedford, T.: *The Warmth Factor in Comfort at Work*. Rep. Ind. Hlth. Res. Bd., London, no. 76, 1936.
2.21 Webb, C. G.: Thermal discomfort in an equatorial climate. *J. Inst. Heat. Vent. Eng.*, Vol. 27 (1960), p. 297.
2.22 Yağlou, C. P. and Minard, D.: Control of heat casualties at military training centres. *AMA Arch. Ind. Hlth*, Vol. 16 (1957), p. 302.
2.23 American Conference of Governmental Industrial Hygienists: *Threshold Limit Values for Chemical Substances and Physical Agents in the Workroom Environment 1980*. ACGIH, Cincinnati, 1980.
2.24 Belding, H. S. and Hatch, T. F.: Index for evaluating heat stress in terms of resulting physiological strains. *Heating, Piping and Air Conditioning*, Vol. 27(8) (1955), p. 129.
2.25 McArdle, B., Dunham, W., Holling, H. E., Ladell, W. S. S., Scott, J. W., Thomson, M. L. and Weiner, J. S.: *The Prediction of the Physiological Effects of Warm and Hot Environments*. Med. Res. Coun., London, RNP Dep no. 47/391, 1947.
2.26 McIntyre, D. A.: *Indoor Climate*. Applied Science, London, 1980.
2.27 Edholm, O. G.: *Man—Hot and Cold*. Edward Arnold, London, 1978.
2.28 Institute of Heating and Ventilation Engineers: *Guide A1: Comfort 1970*. IHVE (now CIBS), London, 1970.
2.29 Yaglou, C. P., Riley, E. C. and Coggins, D. I.: Ventilation requirements. *Trans. Amer. Soc. Heat. Vent. Eng.*, Vol. 42 (1936), p. 133.
2.30 Building Research Establishment: *Digest 206: Ventilation Requirements*. HMSO, London, 1977.
2.31 Loudon, A. G.: *The Effect of Ventilation and Building Design Factors on the Risk of Condensation and Mould Growth in Dwellings*. Building Research Establishment Current Paper CP 31/71, 1971.

2.32 CHARTERED INSTITUTION OF BUILDING SERVICES: *CIBS Guide B2: Ventilation and Air Conditioning Requirements*. CIBS Publications, London, 1972.
2.33 HEALTH AND SAFETY EXECUTIVE: *Guidance Note EH 15/79: Threshold Limit Values 1979*. HMSO, London, 1980.
2.34 DESSAUER, F.: *Zehn fahre Forschung auf dem Physikalische-Medizinischen Grenzegebeit*. Georg Thieme, Leipzig, 1931.
2.35 YAGLOU, C. P., BRANDT, A. D. and BENJAMIN, C. L.: Observations on a group of subjects before, during and after exposure to ionised air. *J. Ind. Hyg.*, Vol. 15 (1933), p. 341.
2.36 HUMPHREYS, C. M. and JENNINGS, B. H.: Atmospheric ions in relation to comfort and other responses of normal individuals. *ASHRAE Journal*, Vol. 4 (1962), p. 55.
2.37 SULMAN, F. G.: Serotonin—migraine in climatic heat stress, its prophylaxis and treatment. *Proceedings of the International Headache Symposium*, Elsinore, Denmark, 1971, p. 205.
2.38 KREUGER, A. P. and REED, E. J.: Biological impact of small air ions. *Science*, Vol. 193 (1976), p. 1209–1213.
2.39 BACH, C.: *Ions for Breathing*, Pergamon, London, 1967.
2.40 JONES, D. P., O'CONNOR, S. A., COLLINS, J. V. and WATSON, B. W.: Effect of long-term ionised air treatment of patients with bronchial asthma. *Thorax*, Vol. 31(4) (1976), p. 428.
2.41 HAWKINS, L. H. and BARKER, T.: Air ions and human performance. *Ergonomics*, Vol. 21(4) (1978), p. 273.
2.42 CHARTERED INSTITUTION OF BUILDING SERVICES: *Proceedings of the Man, Environment and Buildings Symposium*, 28 September 1978, Loughborough. CIBS Publications, London, 1978.
2.43 WORLD HEALTH ORGANISATION: *Proceedings of the WHO First International Indoor Climate Symposium*, Copenhagen, Aug./Sept. 1978. WHO, Geneva, 1978.
2.44 ROYAL COLLEGE OF PHYSICIANS: *Air Pollution and Health*. Pitman, London, 1970.
2.45 BRODINE, V.: *Air Pollution*. Harcourt Brace Jovanovich, New York, 1973.
2.46 STERN, A. C. (Ed.): *Air Pollution*. Academic Press, New York, 2nd edn, 1968.
2.47 MORGAN, G. B., OZOLINS, G. and TABOR, E. C.: Estimated emissions of principal pollutants in the United States during 1968. *Science*, Vol. 170 (1970), p. 289.
2.48 COX, R. A., EGGLETON, E. J., DERWENT, R. G., LOVELOCK, J. E. and PACK, D. H.: Long-range transport of photochemical ozone in north-western Europe. *Nature*, Vol. 255 (1975), p. 118.
2.49 COX, R. A., DERWENT, R. G. and SANDALLS, F. J.: *Some Air Pollution Measurements Made at Harwell, Oxfordshire, During 1973–1975*. HMSO, London, 1976.
2.50 SUMMER, W.: *Odour Pollution of Air: Causes and Control*. Leonard Hill, London, 1971.
2.51 PATTY, F. A. (Ed.): *Industrial Hygiene and Toxicology: Vol. II, Toxicology*. Wiley Interscience, New York, 1963.
2.52 JONASSEN, N. and MCLAUGHLIN, J. P.: Exhalation of radon-222 from building materials and walls. Presented at the *Symposium on the Natural Radiation*

Environment III, Houston, Texas, April 23–28, 1978 (available from the Laboratory of Applied Physics, Techn. University of Denmark, Lyngby).

2.53 ANDERSEN, I. Formaldehyde in the indoor environment—health implications and the setting of standards. *Proceedings of the WHO First International Indoor Climate Symposium*, Copenhagen, Aug./Sept. 1978. WHO, Geneva, 1978.

2.54 LEE, G. L. and SMITH, D. J.: Steelwork insulated with sprayed crocidolite asbestos: controlling a potential hazard. *Ann. Occup. Hyg.*, Vol. 17 (1974), pp. 49–52.

2.55 DUGGAN, M. J. and CULLEY, E. W.: The counting of small numbers of asbestos fibres on membrane filters: a comparison of results from some commercial laboratories. *Ann. Occup. Hyg.*, Vol. 21 (1978), pp. 85–89.

2.56 KINNERSLY, P.: Health hazard guide. *New Civil Engineer* (23rd Nov. 1978), pp. 24–31.

Chapter 3

Steady State Heat Transfer Through Building Fabric

3.1 INTRODUCTION

In the building design process the site, orientation, general form, planning and fabric are all considerations which must be examined, for a multitude of reasons, at the initial stages of design. One of these reasons is that all these aspects have a bearing on the thermal performance and energy consumption of the building in one way or another. This chapter examines the basic thermal performance of buildings through simple considerations of heat losses and gains which, once appreciated, can then be used to assist in making sensible and appropriate thermal decisions relating to these factors.

3.2 FABRIC HEAT TRANSFER

3.2.1 Thermal Transmittance or U-Value

The thermal transmittance or *U*-value is the parameter which gives a measure of the ease with which heat can pass through the fabric of a building. The heat flow processes which occur are conduction through solid material and convection and radiation transfer at surfaces. The concept and means by which *U*-values are calculated have been covered in Section 1.6. Invariably for typical structures they are quoted directly and need not be calculated except for purposes of checking or if an unusual construction is envisaged.

Examples of *U*-values of a variety of wall and roof constructions are given in Tables 3.1 and 3.2. More detailed and extensive values are given elsewhere [3.1, 3.2, 3.3, 3.4, 3.37].

3.2.2 The Effect of Moisture and Exposure on U-Values

The thermal conductivity of porous materials is affected by the moisture content of the material. This is because air pores, which provide the main contribution to the insulation value, are replaced by pores filled with water,

Table 3.1 U-values for a range of wall constructions (adapted from CIBS Guide A3 [3.1])

Basic construction	External finish	Cavity treatment	Internal finish	Total thickness (mm)	U-value (W/m^2 °C) for normal exposure
(1) Solid brickwork	—	—	—	105	3·3
(2) Solid brickwork	—	—	—	220	2·3
(3) Solid brickwork	—	—	—	335	1·7
(4) Solid concrete (cast)	—	—	—	200	3·1
(5) 150 mm concrete block, lightweight	Tile hanging	—	15 mm lightweight plaster	200	0·97
(6) Cavity brickwork, 105 mm brick inner and outer leaves, 50 mm cavity	—	—	15 mm hard plaster	275	1·6
(7) As (6), but with 100 mm lightweight concrete block inner leaf	—	—	15 mm hard plaster	270	0·96
(8) As (7)	—	25 mm expanded polystyrene	15 mm hard plaster	270	0·6
(9) As (7)	—	Cavity filled with 50 mm glass fibre slabs	15 mm hard plaster	270	0·4

(10) Timber frame, 100 mm	Weather-boarding	—	10 mm plasterboard	130	1·8
(11) As (10)	Weather-boarding	50 mm glass fibre	10 mm plasterboard	130	0·6
(12) As (10)	Weather-boarding	100 mm glass fibre	10 mm plasterboard	130	0·3
(13) Corrugated sheeting: 5 mm thick asbestos cement	—	—	—	5	5·3
(14) As (13) with inner lining	—	25 mm cavity	Aluminium foil backed plasterboard	40	1·8
(15) As (13)	—	25 mm glass fibre		40	0·8
(16) Plastic covered steel	—	—	—	5	5·7
(17) As (16) with inner lining	—	25 mm + glass fibre	Aluminium foil backed plasterboard	40	0·9

N.B. Vapour barriers or building paper are not specified in the above constructions as they have negligible influence on the *U*-value. However, they may be required for vapour/moisture control

Table 3.2 U-values for a range of roof constructions (adapted from CIBS Guide A3 [3.1])

Construction	*U*-value (W/m^2 °C) normal exposure
Domestic type pitched roof (1)–(4)	
(1) Slates or tiles on battens, felt backing, plasterboard ceiling	1·9
(2) As (1) but with 50 mm glass fibre quilt insulation between joists	0·54
(3) As (2) but with 100 mm glass fibre	0·30
(4) As (2) but with 200 mm glass fibre	0·16
Large span type low pitched roof (5)–(11)	
(5) Corrugated asbestos cement sheeting	6·1
(6) As (5) but with cavity and aluminium foil backed 10 mm plasterboard lining	1·9
(7) As (6) but with 25 mm glass fibre in cavity	0·94
(8) As (5) with 50 mm glass fibre insulating board fixed below purlins	0·60
(9) Corrugated plastic covered steel sheeting	6·7
(10) Corrugated aluminium sheeting (whilst bright)	3·8
(11) As (9) but with 50 mm glass fibre insulating board fixed below purlins	0·60
Concrete type flat roof (12), (13)	
(12) 20 mm asphalt on 150 mm concrete	3·3
(13) As (12) with ventilated cavity and 50 mm glass fibre and 10 mm plasterboard ceiling	0·53
Woodwool slab type flat roof (14), (15)	
(14) 20 mm asphalt on screed on 50 mm woodwool, with ventilated cavity and 10 mm plasterboard ceiling	1·2
(15) As (14) with 50 mm glass fibre on plasterboard	0·44
Timber joist type flat roof (16)–(18)	
(16) Bituminous felt or boarding on timber joists with 100 mm ventilated cavity and 10 mm plasterboard ceiling	1·9
(17) As (16) with 50 mm glass fibre in cavity	0·51
(18) As (16) with 100 mm glass fibre in cavity	0·26

N.B. Vapour barriers or building paper are not specified in the above constructions as they have negligible influence on the *U*-value. However, they may be required for vapour/moisture control

Table 3.3 Effect of moisture content on the thermal conductivity of brickwork for two density values [3.5]

Density, dry (kg/m^3)	Moisture content (% by volume) 1	2·5	5	10	15	20	25
	Thermal conductivity (W/m °C)						
1 000	0·24	0·29	0·32	0·39	0·43	0·47	0·51
2 000	0·92	1·10	1·24	1·49	1·66	1·80	1·95

which has a higher conductivity value than air. The effects on brickwork are illustrated in Table 3.3.

Hence for walls of porous materials in areas of high exposure to driving rain such changes should be taken into account in quoting the *U*-value of the fabric. The procedure is set out in BRE Digest 108 [3.6] for masonry materials.

The effect is particularly damaging to insulation materials, e.g. glass fibre quilt, which may become moist in roof or wall spaces because of leaks or condensation, and consequently lose a great deal of their insulation value.

The exposure and orientation of an external surface will also affect the surface resistance value, and hence the overall *U*-value (Section 1.6.1). This

Table 3.4 Effect of exposure on external resistance (from [3.1])

Building element	Emissivity of surface	Surface resistance for stated exposure (m^2 °C/W) Sheltered	Normal	Severe
Vertical, i.e. wall	High (most materials)	0·08	0·055	0·03
	Low (metallic finish)	0·11	0·067	0·03
Horizontal, surface up, i.e. roof (more exposed than wall)	High (most materials)	0·07	0·045	0·02
	Low (metallic finish)	0·09	0·053	0·02

Exposure:
- Sheltered: Up to third floor of buildings in city centres
- Normal: Most suburban and country premises: fourth to eighth floors of buildings in city centres
- Severe: Buildings on the coast or on exposed hill sites: floors above the fifth of buildings in suburban or country districts: floors above the ninth of buildings in city centres

Table 3.5 Effect of exposure on the U-value of single glazing (from [3.1])

Type of glazing	U-value (W/m^2 °C)		
	Sheltered	Normal	Severe
Single glazing—vertical	5·0	5·6	6·7
Roof glazing (single)—horizontal	5·7	6·6	7·9

is particularly pertinent to glazing where a large proportion of the U-value is due to the surface resistances. Table 3.4 shows the effect of exposure on external resistance, and Table 3.5 the resultant change to the U-value of vertical and horizontal single glazing. The effect is not so marked on constructions of low U-value because the external surface resistance provides only a small contribution to the overall resistance.

The same effect also occurs on the internal surface of glazing, although to a lesser degree. If, for instance, air is blown over the window surface as part of the heating or ventilation system, this will tend to lower the surface resistance.

3.2.3 U-Value of Floors

The heat loss through a solid ground floor is concentrated around the perimeter of the floor. The ground under the centre of the floor becomes warm, and remains so because of the relative stability of ground temperatures below about 0·5 m. Hence the temperature difference through

Fig. 3.1. Diagrammatic representation of heat loss by conduction through a solid ground floor slab, showing a concentration of loss around the perimeter of the floor (from BRE Digest 145 [3.7]).

the floor at its centre is small, the main drop being concentrated around the perimeter. This is illustrated in Fig. 3.1.

The result of this is that the effective *U*-value of a floor is dependent on the area of the floor. As the floor area increases the loss per unit area decreases. Typical *U*-values are given in Table 3.6. Programmable calculator techniques can be employed to provide ease and flexibility in the determination of these values [3.37].

Table 3.6 U-values of solid floors in contact with the ground, showing an increase in the U-value as the floor size decreases and the number of exposed edges increases (from [3.7])

Floor dimension (m)	*U*-value (W/m^2 °C)	
	Four exposed edges	Two exposed edges at right angles
30 × 30	0·26	0·15
30 × 15	0·36	0·21
30 × 7·5	0·55	0·32
15 × 15	0·45	0·26
15 × 7·5	0·62	0·36
7·5 × 7·5	0·76	0·45
3 × 3	1·47	1·07

Hence solid floors are best insulated by adding insulation material around the perimeter. Very little is to be gained by insulating under the whole slab, except for small floors. The effects of edge insulation are given in Table 3.7.

The same effect applies to any solid floor in contact with the ground. A suspended timber floor, however, usually has a ventilated air space underneath the floor and so the *U*-value of the floor is that of the construction to the ventilated air, although for a large suspended floor the average temperature of this air will be higher than that for outside air, as some warming will occur as it passes through the void depending on the floor size. This is taken into consideration in quoted *U*-values for suspended floors (Table 3.8). Floors which are not in contact with the ground, e.g. cantilevered or intermediate floors, are treated in a similar fashion to other elements except that the *U*-value varies for the direction of heat flow (see Table 3.9) because of the change of surface resistance with orientation (Section 1.6.1) and the effect of heat flow direction on the resistance of air spaces (Section 3.2.5) if these occur in the floor.

Table 3.7 Effects of edge insulation on the U-value of solid floors with four exposed edges [3.1]

Floor dimension (m)	U-value (W/m^2 °C)			
	No insulation	Edge insulation extending to depth of (m)		
		0·25	0·5	1·0
30 × 30	0·26	0·25 (4)	0·23 (12)	0·21 (18)
15 × 15	0·45	0·43 (5)	0·40 (12)	0·36 (20)
7·5 × 7·5	0·76	0·71 (6)	0·65 (15)	0·57 (25)
3 × 3	1·47	1·33 (10)	1·18 (20)	0·96 (35)

Figures in parentheses are the percentage reduction in U-value

3.2.4 U-Value of Windows

The combined internal and external resistances for a vertical sheet of glass for normal exposure conditions is 0·12 + 0·06, i.e. 0·18 m^2 °C/W. The resistance of the glass itself, with a thermal conductivity value of 1·0 W/m °C and thickness 4 mm, is 0·004/1·0, i.e. 0·004 m^2 °C/W, which is only about 2% of the combined surface resistance. In this example the complete resistance of the glazing is 0·184 m^2 °C/W and hence the U-value is 5·4 W/m^2 °C.

Table 3.8 U-values of suspended timber floors directly above the ground. The overall-U-value is dependent on the size of the floor and the floor finish (from [3.7])

Dimension (m)	U-value (W/m^2 °C)		
	Bare or with lino, plastics or rubber finish	Carpet or cork finish	Any surface finish and 25 mm quilt over joists
30 × 15	0·39	0·38	0·30
30 × 7·5	0·57	0·55	0·39
15 × 15	0·45	0·44	0·33
15 × 7·5	0·61	0·59	0·40
7·5 × 7·5	0·68	0·65	0·43
3 × 3	1·05	0·99	0·56

Table 3.9 U-values of intermediate floors. For a given floor construction the U-value is dependent on the direction of heat flow (from [3.1])

Construction	U-value (W/m² °C)	
	Heat flow down	Heat flow up
Timber:		
20 mm on joists with 10 mm plasterboard ceiling	1·4	1·6
Concrete:		
150 mm with 50 mm screed	2·2	2·7
As above with 20 mm wood floor	1·7	2·0
200 mm hollow tile floor	1·6	1·9
As above with 20 mm wood floor	1·3	1·5

This dependence of the U-value on the surface resistances makes it particularly sensitive to changes in the resistance due to exposure or orientation (i.e. vertical or horizontal), while the total surface resistance can be increased by adding more layers of glass as in double or triple glazing, so lowering the U-value.

The frame in which the glass is held affects the U-value of the overall window. Metal frames have poor thermal resistance, timber frames relatively high resistance. The area of the frame material combined with the

Table 3.10 U-values of typical windows (from [3.1])

Window type	Fraction of area occupied by frame (%)	U-value for stated exposure (W/m² °C)		
		Sheltered	Normal	Severe
Single glazing:				
Wood frame	30	3·8	4·3	5·0
Metal frame	20	5·0	5·6	6·7
Double glazing:				
Wood frame	30	2·3	2·5	2·7
Metal frame with thermal break	20	3·0	3·2	3·5

Where the proportion of frame differs appreciably from the above tabulated values, particularly with wood or plastic, the U-values should be calculated (metal members have a U-value similar to glass)

area of glass itself gives the overall U-value for the window. All these factors are summarised by the values given in Table 3.10.

For a solid wall with windows it is often convenient to determine an overall U-value for the total wall. This is done by proportioning the areas and U-values involved, i.e.

$$\text{Overall } U\text{-value} = \frac{\left\{\begin{matrix}\text{Area} \times U\text{-value} \\ \text{for glazing}\end{matrix}\right\} + \left\{\begin{matrix}\text{Area} \times U\text{-value} \\ \text{for solid wall}\end{matrix}\right\}}{\text{Total area of wall and glazing}}$$

For instance a wall of dimensions 10 m × 3 m, and U-value 0·9 W/m^2 °C has four single glazed windows of 1 m × 1 m and U-value 4·3 W/m^2 °C. The overall U-value is given by

$$\begin{aligned}\text{Overall } U\text{-value} &= \frac{4 \times 4{\cdot}3 + 26 \times 0{\cdot}9}{30} \\ &= \frac{17{\cdot}2 + 23{\cdot}4}{30} \\ &= 1{\cdot}35 \text{ W/m}^2\ {}^\circ\text{C}\end{aligned}$$

i.e. the overall U-value is approximately 1·4 W/m^2 °C. The principle can be applied to as many different components as exist in the wall.

3.2.5 Thermal Resistance of Air Spaces

Heat transfer across an air space depends on both radiation and convection. The influence of these effects is summarised as follows:

(a) Radiation transfer will depend on the emissivities and absorptivities of the surfaces involved. These values are typically 0·9 to 0·95 for common building materials resulting in about two-thirds of the total heat being transferred by radiation (see example in Section 1.4.1). However, if the surfaces are of low emissivity, e.g. 0·2 for shiny aluminium foil, then the radiation transfer is considerably reduced. This is done in practice for instance by lining a cavity with aluminium foil, or using foil backed plasterboard with the foil facing into the cavity. (Aluminium foil also acts as a vapour/moisture barrier.)

(b) Natural convection transfer will depend on the width of the cavity and the direction of heat flow.

The resistance of a tall vertical air space increases with its width up to a value of about 20 mm; above this width the resistance is approximately constant.

Natural convection occurs most effectively by upward air flow (Section 1.4.1). Hence the direction of heat transfer across the cavity (e.g. vertically upwards, horizontal or vertically downwards) will affect the resistance value.

(c) Ventilation of the air space will increase heat transfer and lower the effective resistance. The extent will depend on the velocity and temperature of the air and is difficult to quantify in practice.

The effects of these factors are summarised in the resistance values shown for various conditions in Table 3.11.

There are, however, many circumstances where the evaporation and removal of moisture from cavities is of importance to keep the moisture content of the wall materials to appropriately low values to prevent rot and decay. Air flow rates must then be sufficient to ensure that moisture is effectively removed. This evaporation of moisture will have implications for

Table 3.11 Thermal resistance of some examples of air spaces (from [3.1])

Type of air space (unventilated)		Thermal resistance (m^2 °C/W)	
Thickness	*Surface emissivity in space*	*Heat flow horizontal or upwards*	*Heat flow downwards*
5·0 mm	High	0·11	0·11
	Low	0·18	0·18
20 mm or more	High	0·18	0·21
	Low	0·35	1·06
Type of air space (ventilated)		*Heat flow direction dependent on forced air movement*	
Air space behind tiles on tile hung wall		0·12	
Cavity wall construction		0·18	
Air space between asbestos cement cladding and low emissivity surface facing air space (e.g. aluminium foil)		0·3	

the thermal performance of the cavity, although this is difficult to quantify because of the uncertainty of the rates of evaporation likely to be involved.

3.3 PREDICTION OF STEADY STATE HEAT LOSS

3.3.1 Purpose of Calculations

Calculations of heat losses from buildings under steady state conditions (i.e. conditions when temperature and heat flow values remain constant with time) involve two aspects, namely the fabric loss and the ventilation loss. One of the major reasons for performing such calculations is to predict the maximum rate of heat loss, i.e. the heat loss for the lowest temperature conditions frequently encountered externally. This then enables the maximum size and capacity of the building's heating plant to be determined and, for instance, the capacity and number of heat emitters (e.g. radiators) in a particular room to ensure that the room maintains its designed internal temperature during cold weather conditions.

The maximum heat loss values can also be used for comparison purposes as a guide to energy consumption in buildings. For instance if a variety of three-dimensional forms are proposed for a particular building at the design stage, then comparative heat loss figures can assist in selecting minimum heating energy solutions. Such techniques are essentially for guidance only. Comparisons can also be made between the fabric and ventilation losses, and if one is particularly high relative to the other then it is that area where attention should best be concentrated on energy conservation techniques.

Finally, the loss calculations can be applied to estimating seasonal heating requirements, i.e. the total heating energy consumption of the building in a year. One technique is known as the degree day method (Section 3.4) and is useful in estimating seasonal running costs, or comparing the actual running costs of a building (from fuel consumption data) with calculated costs, when assessing the thermal performance of an existing building.

3.3.2 Fabric Heat Transfer

Fabric heat transfer is the flow of heat from one space to another through the solid components of the building fabric from higher to lower temperature regions. Depending on the external weather conditions and required internal conditions heat could be flowing into or out of a building.

The discussion at this stage will be concerned with heat *loss* from a building.

The steady state heat flow process through building fabric has previously been examined and the relationship established in Section 1.6.3, namely

$$\underset{(\text{W})}{\text{Heat flow}} = \underset{(\text{W/m}^2\,^\circ\text{C})}{U\text{-value}} \times \underset{(\text{m}^2)}{\text{Area}} \times \underset{\text{difference }(^\circ\text{C})}{\text{Environmental temperature}}$$

$$Q_f = U \times A \times \Delta t_e$$

To apply the equation, therefore, each fabric component of the building can be taken in turn, its U-value and area established and the heat loss determined knowing the inside and outside environmental temperatures, to give a total fabric heat loss for the building in question. Of the factors involved the areas of the components are easily obtained and the U-values can usually be calculated or taken from standard sources bearing in mind aspects such as moisture content and exposure, as discussed in Section 3.2. The temperatures selected, however, require further consideration.

3.3.3 Selection of Internal and External Environmental Temperatures

The environmental temperature, as described in Section 1.6.2, is given by

$$\begin{matrix}\text{Environmental}\\ \text{temperature}\end{matrix} = \frac{2}{3}\begin{Bmatrix}\text{Mean radiant}\\ \text{temperature}\end{Bmatrix} + \frac{1}{3}\begin{Bmatrix}\text{Air}\\ \text{temperature}\end{Bmatrix}$$

At the design stage of a building, whatever its function, it is usual to specify the internal temperatures required. This may be done either in terms of air and radiant temperatures or else by a proportioning parameter such as the resultant or environmental temperature. Current recommended practice in the UK [3.8] is to specify the resultant temperature, i.e. approximately $\frac{1}{2}$(radiant temperature) + $\frac{1}{2}$(air temperature) as given in Section 2.3.3. The magnitudes of the resultant and environmental temperatures are similar under most conditions in buildings and hence either may usually be used in the fabric heat loss equation. If desired temperature conditions are expressed in other ways it is usually possible to convert these to the environmental temperature. For instance the Fanger comfort graphs (Section 2.3.2) specify mean radiant and air temperatures individually.

It should be noted that for a space with small areas of external wall in a well insulated building the surface (i.e. radiant) and air temperatures will be similar. Hence the environmental temperature approximates to the air temperature.

Thus, a value for the internal environmental temperature can be established for the particular building in question.

The external environmental temperature is affected by both direct solar radiation and radiation losses to the sky, particularly on clear nights. Thus, for example, when solar radiation strikes the external surface of a building, localised surface heating occurs and the heat transfer rate through the fabric is affected. This leads to the definition of the external environmental temperature, or *sol–air temperature* as it is often termed, namely:

> *The sol–air temperature is that outside air temperature which, in the absence of solar radiation, would give the same temperature distribution and rate of heat transfer through the wall or roof as exists with the actual outdoor temperature and incident solar radiation.*

Its calculation is rather more complex than the internal environmental temperature.

However, for heat loss calculations outside conditions are often considered to be overcast, i.e. with no direct solar heating or any marked radiation losses to clear sky, and the external environmental temperature is then approximately equal to the outside air temperature. If there are appreciable losses to the clear night sky, then the external environmental temperature value is lower than the air temperature.

In applying the fabric heat loss equation, particularly for the sizing of heating plant, some value for the severity of the weather conditions prevailing on the site in question must be selected. This is obtained from weather data for the region of the country involved and is typically selected as the air temperature which is exceeded 95 % (or a similar percentile) of the time for an average year. Examples for low altitude sites are typically −19 °C for areas of Sweden, −1 °C for southern England, +3 °C for North Africa and −15°C for northern USA [3.2, 3.10].

The temperature selected also depends on other aspects such as the thermal inertia of the building, whether there are any particularly stringent control requirements, for instance with certain industrial processes, the nature of the heating system and its overload capacity [3.2, 3.10].

In summary, the internal environmental temperature selected is obtained from the specified comfort or design conditions, and in some cases approximates to the internal air temperature. The external environmental temperature, for heat loss determinations, is taken from weather data for the region involved, and may approximate to the external air temperature.

3.3.4 Heat Transfer by Ventilation

Heat is lost from a building when cold air from outside replaces warm air inside, while heat gain occurs if warm external air replaces cooler internal air. The rate of heat transfer depends on the volume of fresh (as opposed to recirculated) air moving through the building in a given time. This volume is determined from the fresh air ventilation rate specified in room volumes per hour (RVH) (see Section 2.4.3) and the volume of the room or building involved. This gives an air volume per hour, which is converted to a volume per second (i.e. divided by 3600) as the heat transfer units are in joules per second, i.e. watts.

The heat carried by this volume depends on the temperature difference between the outgoing and incoming air, i.e. on the *air* temperature difference, not the environmental temperature difference as in the fabric heat transfer equation. It is also dependent on the quantity of heat a given volume of air can carry, that is its volumetric specific heat, which is defined as the quantity of heat required to raise 1 m^3 of air through 1 °C. This value is approximately 1300 J although it varies with the temperature (i.e. density) of the air, atmospheric pressure and relative humidity, and in some cases corrections must therefore be made. Bringing these factors together gives the heat transfer equation for ventilation:

$$\begin{array}{c} \text{Heat} \\ \text{transfer by} \\ \text{ventilation} \\ \text{(W)} \end{array} = \begin{array}{c} \text{Volumetric} \\ \text{specific} \\ \text{heat of air} \\ (\text{J/m}^3\ ^\circ\text{C}) \end{array} \times \begin{array}{c} \text{RVH (fresh air)} \\ \text{converted to} \\ \text{seconds} \\ \ \end{array} \times \begin{array}{c} \text{Volume} \\ \text{of} \\ \text{space} \\ (\text{m}^3) \end{array} \times \begin{array}{c} \text{Air temperature} \\ \text{difference between} \\ \text{inside and outside} \\ (^\circ\text{C}) \end{array}$$

$$Q_v = 1300 \times \frac{\text{RVH}}{3600} \times V \times (t_{ai} - t_{ao})$$

which simplifies to

$$Q_v = 0{\cdot}36 \times \text{RVH} \times V \times \Delta t_a$$

The air temperature difference in the above expression is for heat loss conditions; the temperatures would be reversed for heat gain.

An alternative way of expressing this relationship is in terms of the ventilation rate per person for the occupied space and the numbers of persons involved, i.e.

$$Q_v = 1300 \times vrpp \times P \times \Delta t_a$$

where *vrpp* is the ventilation rate per person in m^3/s and P is the population in the space.

Finally, if the ventilation rate is specified for a space in m^3/s (as might occur for a localised extract system removing pollutants) then the ventilation transfer is

$$Q_v = 1300 \times vr \times \Delta t_a$$

where vr is the ventilation rate in m^3/s.

In applying the above expressions the ventilation rate for the particular building or room (of a given volume) must be known. This is usually specified at the design stage for the activities involved (see Section 2.4.3). However, in naturally ventilated buildings the actual rate occurring will depend on many factors, e.g. site, exposure, wind speed, window design and tightness of construction, and the values can only be estimated (see Section 4.4.1 [3.2, 3.11]). For heat loss determinations this estimate is usually for typical high wind velocity conditions, while for internal heat gain the estimate is taken for still air conditions.

The air temperature difference is also required. The internal air temperature can be obtained from the specified temperature conditions in the space. If this specification is in terms of air and mean radiant temperatures then the former is used directly. However, if the environmental or resultant temperature is specified then the air temperature will depend on the balance occurring between air and mean radiant temperatures. This is primarily dependent on whether the space is heated by warm air (convective) or by radiant means. In the former case the air temperature will be higher than the mean radiant temperature, while with the latter the reverse occurs. For a given environmental temperature the air and mean radiant temperatures depend on the heating type in the following fashion:

(a) For convective heating only the temperatures depend on the insulation of the structure, because this will affect the temperature drop at internal surfaces (Section 1.7.3) and hence the mean radiant temperature; thus the air temperature provided by the heating system must be higher to compensate for this.

(b) For radiant heating the temperatures depend mainly on the ventilation loss alone, because this will affect the air temperature, which will be compensated for by a higher mean radiant temperature, provided by the heating system.

Table 3.12 Air and mean radiant temperatures for a given environmental temperature in a space for: (a) mainly convective heating (irrespective of rate of ventilation)

Enclosure		Air temperature (in °C) corresponding to the following environmental temperatures for an outdoor air/environmental temperature of −1 °C								
		15			18			21		
		With average U-value of exposed surfaces (wall, windows, etc.) (in W/m^2 °C)								
Type	Exposed surfaces (excluding floor)	0·6	1·8	3·0	0·6	1·8	3·0	0·6	1·8	3·0
Large enclosure,	Roof and 4 walls	16·4	19·2	22·0	19·6	22·8	26·0	23·9	29·7	35·5
e.g. factory	Roof and 2 walls	16·0	18·0	20·0	19·2	21·6	24·0	22·4	25·2	28·0
30 m × 30 m × 9 m	1 wall	15·2	15·6	15·9	18·2	18·7	19·1	21·3	21·8	22·3
Large open floor,	Roof and 4 walls	16·2	18·6	21·0	19·5	22·5	25·5	22·7	26·1	29·5
e.g. office	4 walls	15·4	16·2	17·0	18·4	19·2	20·0	21·4	22·2	23·0
30 m × 30 m × 3 m	2 walls	15·2	15·6	16·0	18·2	18·6	19·0	21·2	21·6	22·0
Small room,	Roof and 4 walls	16·5	19·6	22·7	19·8	23·5	27·2	23·1	27·4	31·6
e.g. domestic or	4 walls	16·1	18·3	20·5	19·3	21·9	24·5	22·5	25·5	28·5
office	2 walls	15·5	16·6	17·7	18·6	19·9	21·2	21·8	23·3	24·8
5 m × 5 m × 3 m	1 wall	15·3	15·8	16·3	18·3	18·9	19·6	21·4	22·1	22·9

Table 3.12—contd.
(*b*) *mainly radiant heating* (*irrespective of average U-value of enclosing surfaces*) (*adapted from* [*3.12*])

Enclosure: Type	Enclosure: Ventilation allowance (W/m^3 °C)	Air temperature (in °C) corresponding to the following environmental temperatures for an outdoor air/environmental temperature of −1 °C: 15	18	21
Large enclosure, e.g. factory 30 m × 30 m × 9 m	0·33	12·4	14·9	17·4
	0·66	10·5	12·7	14·8
Large open floor, e.g. office 30 m × 30 m × 3 m	0·33	13·7	16·5	19·2
	0·66	12·6	15·2	17·7
Small room, e.g. domestic or office 5 m × 5 m × 3 m	0·33	14·2	17·1	20·0
	0·66	13·6	16·4	19·1

Note: A ventilation allowance of 0·33 W/m^3 °C is approximately equivalent to a fresh air ventilation rate of 1 room volume per hour, and 0·66 W/m^3 °C to two room volumes per hour

Techniques are available for estimating air temperatures for the two circumstances (the 'energy balance triangle' approach) [3.9], and summary resulting data for certain building types is given in Table 3.12.

Typically, in a large factory with two to four exposed walls and an exposed roof with an overall U-value of 1·8 W/m^2 °C, if the environmental temperature is 18 °C and the outside temperature −1 °C then for mainly convective heating the air temperature will be 22 °C. For mainly radiant heating the air temperature will be about 15 °C for a fresh air ventilation rate of about 1 RVH, and 13 °C if the ventilation rate is doubled. In a poorly insulated small room (U-value of approximately 3·0 W/m^2 °C) and all surfaces part of the external fabric, then for an internal environmental temperature of 21 °C and outside temperature of −1 °C, for mainly convective heating the air temperature would need to be over 30 °C to compensate for the cold walls, while for a well insulated room with only one wall exposed the air, environmental and mean radiant temperatures are approximately equal.

In general, for convective heating, as the U-value and area of exposed wall in the space increases, so the air temperature must increase to maintain the specified environmental temperature (because the mean radiant temperature is decreasing). For radiant heating as the ventilation rate increases, so the air temperature decreases, but must be compensated for by increased radiant heating to maintain the desired environmental temperature.

Thus the form of heating will affect the ventilation heat loss, even though the specified environmental temperature and ventilation rate remain constant.

However, in the case of convective heating, as the insulation value of the fabric increases, so the air and radiant temperatures become closer in value and the air temperature approximates to the environmental temperature, and therefore either may be used in the ventilation loss equation.

For heat loss calculations the external air temperature is obtained from weather data in the same manner as for fabric heat loss and often the same 'average worst' air temperature is used. For summer heat gain a similarly selected high external air temperature value is used.

3.3.5 Example of Design Heat Loss Calculation

A detached house in a temperate region on a site of normal exposure is considered. The dimensions of the house are 6 m × 10 m × 6 m (high) with a total of 35 m^2 glazing. The fabric components are as follows:

Component	Construction	Area (m^2)	U-value ($W/m^2\ °C$)
Walls	105 mm brick/ 25 mm cavity/ 25 mm expanded polystyrene/100 mm aerated concrete block/15 mm plaster	Total walls minus windows = 157	0·6
Windows	Double glazing in aluminium frames	20 downstairs 15 upstairs	3·2
Floor	Suspended timber with carpets	60	0·6
Roof	Pitched tile, with 80 mm glass fibre quilt above ceiling	Taken as ceiling, 60	0·4

Design temperatures. The recommended dry resultant temperatures for dwellings (where occupants are assumed to be wearing light to medium clothing and are sedentary or participating in light activity) are given in the CIBS Guide A1 [3.8] as

Living areas, i.e. downstairs	21 °C
Bedrooms, i.e. upstairs	18 °C

in this example.

There are slight differences for recommended temperatures for areas such as entrance halls, stairways and bathrooms, but in this example the dwelling will be treated as the two areas defined above.

The dry resultant temperature can be assumed to be equal to the environmental temperature (Section 3.3.3). The heating system is to be mainly natural convective. The temperature components therefore become, to the nearest °C, interpreting from Table 3.12(a):

	Dry resultant temperature	Air temperature, t_a	Mean radiant temperature, t_r	Environmental temperature, t_e
Downstairs	21	22	20	21
Upstairs	18	19	17	18

The above temperature values could also be obtained from Fanger or standard effective temperature approaches although the actual values recommended for comfort may vary slightly (Section 2.3.6).

These temperature values are averages for the space. There will be localised variations, e.g. with height and distance from heat emitters. There will also be some variations from room to room, as previously mentioned. This can be taken into account in room by room calculations (as demonstrated after the overall calculation).

It should be noted that in this example there is little difference between the air and mean radiant temperatures. Approximations and simplifications that can be made will be discussed at the end of the example, but at this stage the full procedure for the heat loss determination will be demonstrated.

The external design temperature for the temperate region involved is taken as an air and environmental temperature of −1 °C.

For the few occasions when conditions fall below this value, and internal temperatures cannot be maintained by the designed heating system, it is assumed that occupants will take appropriate measures to maintain comfort, e.g. through additional clothing or extra heating appliances.

Hence the required temperature differences for the calculations are:

	For fabric loss: Environmental temperature difference, Δt_e (°C)	For ventilation loss: Air temperature difference, Δt_a (°C)
Downstairs	22	23
Upstairs	19	20

Fabric heat loss determination

Component	Heat loss (W) $Q_f = U \times A \times \Delta t_e$
Solid walls	0·6 × 157 × 22 = 2 072
Downstairs windows	3·2 × 20 × 22 = 1 408
Floor	0·6 × 60 × 22 = 792
Upstairs windows	3·2 × 15 × 19 = 912
Roof	0·4 × 60 × 19 = 456
	Total 5 640 W

The fabric loss is 5·6 kW.

Ventilation heat loss determination. The maximum ventilation rate (RVH) occurring, with doors and windows shut and high external wind speeds, is assumed to be 1 per hour throughout the house. The volume is considered simply as the total volume of the envelope, without any allowance made for internal walls, floors or furnishings. This may be acceptable in a house, but for instance in a warehouse the volume involved may be altered considerably by the volume of stored products.

Space	Ventilation loss (W) $Q_v = 0{\cdot}36 \times \text{RVH} \times V \times \Delta t_a$
Downstairs	$0{\cdot}36 \times 1 \times 180 \times 23 = 1\,490$
Upstairs	$0{\cdot}36 \times 1 \times 180 \times 20 = 1\,296$
	Total 2 786

The total ventilation loss is 2·8 kW.

$$\text{Fabric} + \text{ventilation loss} = 5{\cdot}6 + 2{\cdot}8 = 8{\cdot}4\,\text{kW, i.e.}$$

$$\textit{Total design heat loss} = 8{\cdot}4\,\text{kW}$$

Internal heat gains (Section 3.5) have not been considered in the example as the design heat loss procedure is considering the worst conditions, i.e. when there are no contributing heat gains to counteract heat losses.

Sizing of heating plant. If the building is to be heated from a centralised plant then the total design heat loss figure is used to size this plant. Usually a safety margin is added to the calculated loss. The margin depends on the accuracy of the data and calculations involved, which for sophisticated analyses can be very high (see Section 6.3). However, in the steady state approach illustrated here it is usually recommended to make a 20–30 % addition to the calculated loss. This is to compensate for factors such as:

(a) Lack of confidence in U-values, especially at construction junctions.
(b) Lack of confidence in ventilation rates.
(c) Heat loss from pipes or ducts in areas where the loss is not contributing to useful heating.
(d) If the heating system is used intermittently then further margins may be added to provide appropriately rapid warm-up times for plant and building (see Section 6.5).

Hence with the safety margin addition the heating plant should have a rated output of 11 kW.

If hot water is required from this plant then a further addition should be made, which can be estimated from standard sources (e.g. [3.13]). Typically in this case 3 kW would be added making the total output 14 kW. The final choice of heating plant size may be influenced by the actual sizes available from manufacturers.

Room heat emitters. To size room heat emitters, whether they be supplied from a centralised heating plant or are independent units, the heat loss calculation method is repeated, but on a room by room basis, as illustrated below for one room from the worked example.

The downstairs living room has dimensions of approximately 6 m × 3 m × 2·4 m high with 10 m^2 of glazing. One 6 m length wall is internal. In the calculation, heat loss to the upstairs through the intermediate floor is taken into account, although its contribution is small. Ventilation heat loss to the upstairs is not considered as it is assumed that there are intermediate doors which would be shut in cold weather conditions. However, in an open stairway design an additional estimated allowance would need to be made.

Fabric loss

Component	Area (m^2)	U-value (W/m^2 (°C)	Δt_e (°C)	Fabric loss (W)
Solid walls	18·8	0·6	22	248
Windows	10	3·2	22	704
Floor	18	0·6	22	238
Ceiling (intermediate floor)	18	1·6	3	86
				1 276

Ventilation loss

Volume (m^3)	RVH	Δt_a (°C)	Loss (W)
43·2	1	23	358

The total loss from the room is 1276 + 358 = 1634 W. A 30 % safety margin raises this to 2100 W.

Hence the heat emitter(s) in the room should have a rated output of 2·1 kW.

3.3.6 Approximations and Simplifications in the Design Calculation

The example has demonstrated the relationship and use of dry resultant, environmental, air and radiant temperatures. However, if the recommended dry resultant temperatures were used to establish the temperature differences throughout the calculations, without separating out the air and radiant components, the only change would be a reduction in the ventilation loss from 2·8 kW to 2·7 kW. This is a small change in the light of the 25–30 % safety margin added at the end of the calculation. Therefore it would be acceptable in this example to use a single design temperature throughout, as recommended in some design guides [3.3, 3.4, 3.14]. Whether this temperature value corresponds to the dry resultant, environmental or air temperature is of little importance in terms of the overall accuracy of the calculation.

However, such approximations should be treated with care for in certain examples large errors would ensue. For instance, in a poorly insulated dwelling, or one with large areas of single glazing, in cold conditions a dry resultant (or environmental) temperature of 21 °C would correspond to a radiant (i.e. average surface temperature) of approximately 17 °C and an air temperature of 25 °C, producing about a 15 % increase in the ventilation loss.

Alternatively in a factory with mainly radiant heating a resultant temperature of 18 °C is equivalent to an air temperature of 15 °C and hence a radiant temperature of 21 °C (see Table 3.12(b)), while for a swimming pool with large areas of glazing the mean radiant temperature will be considerably lower than the air temperature.

In such cases there are two alternatives: (a) to make estimates of the likely temperatures occurring, either by using, for instance, Table 3.12, or estimating surface temperatures for given fabric *U*-values (see Section 1.7.3); (b) to take the 'energy balance triangle' calculation approach [3.9, 3.15].

3.3.7 Variations on the Design Example

The example can be used to demonstrate the effects on rated heating plant output of alterations in the fabric and heating systems.

(*a*) *Fabric of poorer insulation value and ventilation control.* The original example is taken with fabric alterations as below:

Walls: The insulation in the cavity is not included, raising the *U*-value from 0·6 to 1·0 W/m² °C.

Windows: These are single rather than double glazed, i.e. the *U*-value changes from 3·2 to 5·6 W/m² °C.

The single glazing is assumed to allow greater air leakage, causing a doubling of the ventilation rate to 2 RVH.

On repeating the calculations the total fabric loss rises from 5·6 kW to 8·8 kW, i.e. approximately a 60 % increase.

The ventilation loss doubles from 2·8 to 5·6 kW.

The total heat loss is now 8·8 + 5·6 = 14·4 kW compared with 8·4 kW originally, i.e. approximately a 70 % increase. This has implications for both the size and capital cost of the heating plant and for running costs.

(*b*) *Mainly radiant rather than convective heating* (e.g. radiant ceiling panels). The original example is taken, the design dry resultant temperatures remaining unchanged, but the air temperature values are lowered (see Table 3.12(b)) because of the change in heating system.

	Old t_a (°C)	New t_a (°C)
Downstairs	22	20
Upstairs	19	17

Consequently, the ventilation loss is altered from 2·8 kW down to 2·5 kW, i.e. about a 10 % decrease (corresponding to the 10 % decrease in air temperature).

In a dwelling, the design implications of radiant heating systems, such as installation and maintenance problems, tend to limit their use, but in a factory, for example, there may be many design advantages together with the reduced heating costs particularly as higher ventilation rates are likely (for further discussion see Section 5.3.3).

3.4 SEASONAL HEATING REQUIREMENTS: THE DEGREE DAY METHOD

Energy and running cost considerations have made it important to be able to assess the overall heat consumption of a building throughout the heating season. The degree day method described by Billington in 1966 [3.16] is a simple technique based on the steady state approach which enables this to be estimated.

Degree days are the difference between a fixed base temperature and the average external air temperature over a period of 24 h. In the UK the base temperature is usually taken as 15·5 °C. Thus if the external temperature on a particular day averages 5 °C then the day is equivalent to 15·5 − 5 = 10·5 degree days. This base temperature value is selected because in typical buildings if the heating system can itself maintain an internal temperature of 15·5 °C then the usual internal heat gains, such as people and lights, bring the actual temperature to approximately 18 to 20 °C, i.e. typical comfort conditions. Days with average temperatures higher than 15·5 °C are discarded. In the USA the base temperature is usually taken as 65 °F (18·3 °C) [3.2]. Degree day data are usually collated by the month or year, for given regions of a country. Typical values for the UK, averaged over 20 years, are given in Table 3.13. Values for any individual month or year may vary considerably from these averaged values, as demonstrated in the table.

Corrections must be made for altitude, namely for each 100 m increase in altitude the overall figure is increased by 100 degree days. Current and averaged degree day data are collated in the UK by the Department of Energy, and appear in a number of current journals [3.18, 3.19], while in the USA data appear in ASHRAE documentation [3.2].

The determination of seasonal heating loads from degree day data is a straightforward procedure, using the heat loss equation already established. The relevant time period in the year must be considered. In some cases the 'heating season' (e.g. from October to April) is the only time during which heating is required; alternatively the whole year may be considered.

3.4.1 Example of Degree Day Calculation

The house taken in the previous example is considered. For convenience the overall U-value for the building is determined by taking the area proportion of each component with its U-value (as described for windows in Section 3.2.4).

$$\text{Overall } U\text{-value} = \frac{U_1A_1 + U_2A_2 + \cdots}{A_1 + A_2 + \cdots}$$

Therefore

$$\text{Overall } U\text{-value} = \frac{0{\cdot}6 \times 157 \text{ (walls)} + 3{\cdot}2 \times 35 \text{ (windows)} + 0{\cdot}6 \times 60 \text{ (floor)} + 0{\cdot}4 \times 60 \text{ (roof)}}{157 + 35 + 60 + 60}$$

$$= \frac{94{\cdot}2 + 112 + 36 + 24}{312}$$

$$= \frac{266}{312} = 0{\cdot}85 \text{ W/m}^2\ {}^\circ\text{C}$$

i.e. the overall U-value of the building is 0·85 W/m^2 °C.

Table 3.13 Degree day data for different regions of the UK for base temperature of 15·5 °C. The first row gives averaged figures over 20 years from 1958 to 1977. The second row contains individual monthly values for the period September 1977 to August 1978. (*Source: Department of Energy, UK* [3.17] *or CIBS Journal* [3.18].)

	Thames Valley	South Eastern	Southern	South Western	Severn Valley	Midland	West Pennines	North Western	Borders	North Eastern	East Pennines	East Anglia	West Scotland	East Scotland	North East Scotland	Wales	Northern Ireland
Sept.	55	84	77	56	69	92	78	94	107	87	77	73	104	106	123	73	100
	61	89	87	65	77	91	87	103	107	98	88	77	116	119	133	86	114
Oct.	132	162	145	115	144	172	155	168	183	170	157	153	179	185	199	138	172
	97	125	107	77	114	143	120	115	153	144	142	112	135	150	160	107	133
Nov.	256	281	253	215	259	290	280	296	300	296	283	283	303	309	322	239	288
	241	263	252	219	260	286	295	304	288	280	273	266	331	317	330	258	323
Dec.	336	359	331	279	332	362	350	361	364	364	354	363	359	371	385	305	347
	273	291	264	229	274	298	291	312	301	309	313	288	320	319	324	269	297
Jan.	349	371	342	296	348	375	363	371	380	379	367	383	373	383	399	327	363
	367	393	354	303	353	391	376	395	396	409	395	391	413	423	426	339	382
Feb.	304	326	307	272	310	335	323	333	343	335	324	336	336	344	360	301	326
	337	363	345	298	343	375	351	377	399	411	389	375	403	428	434	335	349
Mar.	285	307	297	270	294	322	308	321	334	320	307	319	319	328	348	296	314
	252	284	270	252	261	278	275	289	293	276	271	277	297	299	313	279	299
Apr.	199	225	216	201	211	235	225	241	261	235	218	234	238	254	272	231	239
	246	271	260	238	264	285	258	265	305	284	275	279	279	302	309	269	273
May	113	146	141	131	130	153	138	163	193	155	139	143	163	190	206	157	168
	117	144	149	116	129	160	125	138	196	156	147	158	145	178	187	155	153
June	49	75	69	60	58	78	67	83	108	78	67	73	84	102	112	83	86
	56	75	77	62	68	86	88	96	114	91	83	74	90	104	118	98	102
July	24	44	41	32	31	47	41	55	74	49	40	44	61	70	85	49	61
	29	47	44	43	39	57	61	65	95	61	64	56	74	82	101	72	73
Aug.	26	48	42	31	34	51	42	57	73	50	41	43	64	72	86	43	64
	30	54	47	34	36	52	42	56	72	51	48	53	58	58	80	59	61
Totals	2 128	2 428	2 266	1 958	2 220	2 512	2 370	2 543	2 720	2 518	2 374	2 447	2 583	2 714	2 897	2 242	2 528
	2 106	2 359	2 256	1 936	2 218	2 502	2 369	2 515	2 719	2 570	2 488	2 406	2 661	2 779	2 915	2 326	2 559

The house is in the Southern region of the UK. The heating load for the total year is to be estimated.

Fabric loss.

$$Q_f = UAt_{dg}$$

where t_{dg} is the number of degree days involved, converted to seconds. The 20 year degree day data for the Southern region gives 2266 degree days for the whole year which must be multiplied by 24 × 3600 to bring to seconds:

Therefore

$$Q_f = 0{\cdot}85 \times 312 \times 2266 \times 24 \times 3600$$
$$= 5{\cdot}19 \times 10^{10}\,\mathrm{J}$$

i.e.

$$\simeq 52\,\mathrm{GJ} \qquad (1\,\mathrm{GJ} = 10^9\,\mathrm{J})$$

Ventilation loss.

$$Q_v = 0{\cdot}36 \times V \times \mathrm{RVH} \times t_{dg}$$

Assume that the average RVH during heating is 1 (the actual ventilation rate for a particular dwelling would be very much an estimate). Then

$$Q_v = 0{\cdot}36 \times 360 \times 1 \times 2266 \times 24 \times 3600$$
$$= 2{\cdot}54 \times 10^{10}\,\mathrm{J}$$

i.e.

$$\simeq 25\,\mathrm{GJ}$$

The total space heating requirement for the house per year is 77 GJ. The actual fuel used to achieve this heat input must take into account the efficiency of the heating plant.

Some of the assumptions and limitations in this example are given below.

(a) No account is taken of variations of temperature in the house or of differences in air and radiant temperatures.
(b) The heating was considered to be required at all times of the year. The actual 'heating season' would vary for different households.
(c) The heating was considered to be running whenever the external temperature fell below the base temperature of 15·5 °C, both night and day. Systems are usually switched off at night but the extent varies with household.

(d) The U-values of all elements are considered to be constant. But, for instance, drawing curtains over windows lowers their U-value to values which depend on the thickness and fitting of the curtain.

(e) The average ventilation rate for the house is very much an estimate and will depend on the nature of the window and door openings, and occupants' behaviour.

(f) If a variety of fuels are used for heating, e.g. gas and electricity, then the proportioning of these fuels is difficult to assess.

(g) The degree days taken in this example are values averaged over 20 years. For any individual year there may be considerable differences in predicted energy consumption (see Table 3.13).

The above illustrates some of the limitations of the degree day method and the estimates of energy consumption are usually quoted as $\pm 25\%$ accuracy. The technique and some of the refinements and corrections that can be taken are further discussed in Section 6.2 and elsewhere [3.20].

3.5 HEAT GAINS IN BUILDINGS

Heat gains contribute towards heating buildings during the cold season, and could be included in the assessment of heating loads, whereas during warm seasons they can often lead to undesirably high temperature conditions within the building.

Heat gains arise from two sources. There are those internal gains produced from activity and equipment inside the building, known as casual gains. Secondly there are those sources arising from the outside climatic conditions leading to heat flowing into the building through the fabric. Usually this is a direct solar effect caused by heating of the external fabric or by direct radiation transfer through windows. Overheating is exacerbated by simultaneous high external air temperatures which can also mean that ventilation transfers unwanted heat into the building.

3.5.1 Casual Heat Gains

The most usual sources of casual internal heat gains are people, lighting and machinery or equipment.

People. For typical internal activities, for instance office work, each person is considered to act as a continuous heat source giving off approximately 120 W, while for manual activities 220 W is more appropriate (Section 2.2 gives further details). In a densely occupied building, e.g. a theatre, the total contribution from people can be considerable.

Lighting. Only a small fraction of the power given off from a light fitting is in the form of visible light energy. The remainder is heat energy (radiant and convective). The total heat energy amounts to the total rating of the light fittings since the light is converted to heat when it is absorbed by surfaces. Such lighting gains become particularly relevant in deep plan buildings where artificial lighting is operating continuously during occupancy.

Machinery or equipment. In factories and workshops heat gain from mechanical equipment is of importance. The power rating of the equipment will give the heat gain quantities involved. Similarly in offices typewriters and other equipment can make significant contributions to heat gains. In some buildings such as in steel works, the heat gains involved are often so great that excessive temperatures cannot be avoided.

The total heat gain from internal sources can thus be determined, although if some aspects are being handled separately by the mechanical systems (e.g. local extraction over light fittings or hot equipment such as cookers) then these will not contribute directly to the heat gain in the occupied zones.

In heat loss determinations (Section 3.3.5) the total heat gains can be subtracted from the calculated losses to give a net heat input required to maintain internal temperatures. It should be noted, however, that when buildings are being heated initially (e.g. in the mornings before occupancy) then the majority of the gains are not occurring and the heating plant must be rated to cope with this.

3.5.2 Heat Gains through the Fabric

In the simplest analysis heat gains through the fabric are those created instantaneously by direct solar radiation through glazing. This direct radiation falls on to objects and surfaces in a room, causing them to heat up and re-emit heat by radiation and convection (see the greenhouse effect, Section 1.3.6). The amount of energy concerned can be several hundred watts per square metre of glazing. To estimate this heat gain the area and orientation of the glazing must be known, and solar data for the latitude concerned must be available. These data are in the form of the intensity of energy falling on a vertical or horizontal surface for a given time, season and orientation as shown in Tables 3.14 and 3.15 for two latitudes for comparison purposes. It is seen, for instance, that the maximum solar intensity on a south facing vertical window occurs in February/October at 50 °N (645 W/m^2) and in December for 10 °N (485 W/m^2).

The fraction of radiation penetrating through a window depends on the

Table 3.14 Direct and diffuse solar intensities on vertical surfaces and total intensities on a horizontal surface in W/m² for latitude 50° N (derived from [3.21])

Date	Orientation	Daily mean	Sun time																		
			03	04	05	06	07	08	09	10	11	12	13	14	15	16	17	18	19	20	21
June 21	N	30		15	150	145	55	0	0	0	0	0	0	0	0	0	55	145	150	15	
	E	135		20	310	530	635	635	550	400	210	0	0	0	0	0	0	0	0	0	
	S	95		0	0	0	0	75	205	310	375	400	375	310	205	75	0	0	0	0	
	W	135		0	0	0	0	0	0	0	0	0	210	400	550	635	635	530	310	20	
	Diff.	25		0	20	30	35	40	45	50	50	50	50	50	45	40	35	30	20	0	
	Horiz.	345		0	90	235	400	565	700	810	880	910	880	810	700	565	400	235	90	0	
July 23 and May 21	N	20			110	120	25	0	0	0	0	0	0	0	0	0	25	120	110		
	E	130			245	495	625	635	555	405	215	0	0	0	0	0	0	0	0		
	S	110			0	0	0	105	235	345	415	440	415	345	235	105	0	0	0		
	W	130			0	0	0	0	0	0	0	0	215	405	555	635	625	495	245		
	Diff.	25			15	25	35	40	45	50	50	50	50	50	45	40	35	25	15		
	Horiz.	320			55	195	365	525	670	780	850	875	850	780	670	525	365	195	55		
August 24 and April 20	N	5				50	0	0	0	0	0	0	0	0	0	0	0	50			
	E	115				355	565	615	555	410	215	0	0	0	0	0	0	0			
	S	145				0	35	175	315	430	505	530	505	430	315	175	35	0			
	W	115				0	0	0	0	0	0	0	215	410	555	615	565	355			
	Diff.	20				20	30	35	40	45	50	50	50	45	40	35	30	20			
	Horiz.	265				100	250	420	570	680	755	775	755	680	570	420	250	100			
September 22 and March 22	N	0				0	0	0	0	0	0	0	0	0	0	0	0	0			
	E	85				0	360	515	505	390	210	0	0	0	0	0	0	0			
	S	175				0	75	230	390	515	595	625	595	515	390	230	75	0			
	W	85				0	0	0	0	0	0	0	210	390	505	515	360	0			
	Diff.	15				0	20	30	35	40	40	45	40	40	35	30	20	0			
	Horiz.	180				0	100	250	395	515	585	610	585	515	395	250	100	0			

Table 3.14—contd.

Date	Orientation	Daily mean	Sun time																		
			03	04	05	06	07	08	09	10	11	12	13	14	15	16	17	18	19	20	21
October 23	N	0					0	0	0	0	0	0	0	0	0	0	0				
and	E	55					50	325	395	330	185	0	0	0	0	0	0				
February 20	S	170					15	190	375	520	615	645	615	520	375	190	15				
	W	55					0	0	0	0	0	0	185	330	395	325	50				
	Diff.	10					0	20	30	35	35	35	35	35	30	20	0				
	Horiz.	110					5	105	235	340	410	435	410	340	235	105	5				
November 21	N	0						0	0	0	0	0	0	0	0	0					
and	E	30						85	245	245	145	0	0	0	0	0					
January 21	S	135						60	270	440	545	580	545	440	270	60					
	W	30						0	0	0	0	0	145	245	245	85					
	Diff.	10						5	20	25	30	30	30	25	20	5					
	Horiz.	55						15	100	185	250	275	250	185	100	15					
December 22	N	0							0	0	0	0	0	0	0						
	E	20							175	200	125	0	0	0	0						
	S	110							205	380	490	530	490	380	205						
	W	20							0	0	0	0	125	200	175						
	Diff.	5							15	25	25	30	25	25	15						
	Horiz.	45							55	135	195	210	195	135	55						

For southern latitudes, this table may be used by reading northern values for southern aspects and vice versa, substituting dates as follows:

North	June	May July	April August	March September	February October	January November	December
South	December	November January	October February	September March	August April	July May	June

Table 3.15 Solar data as specified in Table 3.14 for latitude 10° N (derived from [3.21])

Date	Orientation	Daily mean	Sun time																		
			03	04	05	06	07	08	09	10	11	12	13	14	15	16	17	18	19	20	21
June 21	N	110				70	200	235	235	225	220	215	220	225	235	235	200	70			
	E	100				160	510	605	545	405	215	0	0	0	0	0	0	0			
	S	0				0	0	0	0	0	0	0	0	0	0	0	0	0			
	W	100				0	0	0	0	0	0	0	215	405	545	605	510	160			
	Diff.	20				10	30	40	45	50	55	55	55	50	45	40	30	10			
	Horiz.	310				25	235	470	685	855	965	1 010	965	855	685	470	235	25			
July 23 and May 21	N	85				55	170	200	190	180	170	165	170	180	190	200	170	55			
	E	105				145	515	615	560	415	225	0	0	0	0	0	0	0			
	S	0				0	0	0	0	0	0	0	0	0	0	0	0	0			
	W	105				0	0	0	0	0	0	0	225	415	560	615	515	145			
	Diff.	20				5	30	40	45	50	55	55	55	50	45	40	30	5			
	Horiz.	315				25	225	470	690	865	985	1 020	985	865	690	470	225	25			
August 24 and April 20	N	30				20	90	90	70	50	40	30	40	50	70	90	90	20			
	E	105				95	520	640	585	435	235	0	0	0	0	0	0	0			
	S	0				0	0	0	0	0	0	0	0	0	0	0	0	0			
	W	105				0	0	0	0	0	0	0	235	435	585	640	520	95			
	Diff.	20				5	30	40	45	50	55	55	55	50	45	40	30	5			
	Horiz.	315				15	215	465	695	875	1 000	1 025	1 000	875	695	465	215	15			
September 22 and March 22	N	0					0	0	0	0	0	0	0	0	0	0	0				
	E	100					495	640	590	440	240	0	0	0	0	0	0				
	S	45					25	65	105	135	155	160	155	135	105	65	25				
	W	100					0	0	0	0	0	0	240	440	590	640	495				
	Diff.	20					25	40	45	50	55	55	55	50	45	40	25				
	Horiz.	305					180	440	670	855	985	1 020	985	855	670	440	180				

Table 3.15—contd.

Date	Orientation	Daily mean	Sun time																		
			03	04	05	06	07	08	09	10	11	12	13	14	15	16	17	18	19	20	21
October 23	N	0					0	0	0	0	0	0	0	0	0	0	0				
and	E	95					435	605	570	425	230	0	0	0	0	0	0				
February 20	S	110					105	195	250	290	315	325	315	290	250	195	105				
	W	95					0	0	0	0	0	0	230	425	570	605	435				
	Diff.	20					25	35	45	50	50	55	50	50	45	35	25				
	Horiz.	280					145	390	620	800	915	955	915	800	620	390	145				
November 21	N	0					0	0	0	0	0	0	0	0	0	0	0				
and	E	85					360	550	530	400	215	0	0	0	0	0	0				
January 21	S	155					150	285	360	405	435	445	435	405	360	285	150				
	W	85					0	0	0	0	0	0	215	400	530	550	360				
	Diff.	20					20	35	40	45	50	50	50	45	40	35	20				
	Horiz.	250					110	340	555	725	835	870	835	725	555	340	110				
December 22	N	0					0	0	0	0	0	0	0	0	0	0	0				
	E	80					330	520	510	385	205	0	0	0	0	0	0				
	S	170					160	310	395	445	475	485	475	445	395	310	160				
	W	80					0	0	0	0	0	0	205	385	510	520	330				
	Diff.	20					20	35	40	45	50	50	50	45	40	35	20				
	Horiz.	235					100	315	525	690	795	835	795	690	525	315	100				

For southern latitudes, this table may be used by reading northern values for southern aspects and vice versa, substituting dates as follows:

North	June	May July	April August	March September	February October	January November	December
South	December	November January	October February	September March	August April	July May	June

orientation and nature of the glazing and influences due to blinds or other solar control devices. This can be determined approximately by multiplying the incident radiation by the 'solar gain factor' for the windows involved, as given in Table 3.16.

With these data the heat gain into the building for particular glazing and time of year can be estimated, and this gain then regarded as an internal

Table 3.16 Solar gain factors for some window types (from [3.22])

Window construction	Solar gain factor
Single 4 mm clear glass	0·77
Double 6 mm clear glass	0·61
Single 4 mm clear glass with internal venetian blind	0·46
Double 6 mm clear glass with white venetian blind between glass	0·28
Single 4 mm clear glass with external canvas roller blind	0·11

heat source which may be contributing to heating the building in cold seasons, or may need to be removed for comfort to be maintained in warm weather.

This procedure does not, however, take any account of heat stored by direct solar gain in the internal fabric, nor heat transfer from outside to inside through the solid fabric. These aspects are treated in Section 5.2.6.

3.5.3 Control of Heat Gains

The problems of heat gain and resulting high temperatures in buildings are best approached by controlling the gains at source rather than dealing with them when the problem has affected other spaces in the building. For example, heat can be removed directly from light fittings from above and around the fitting by ventilation. This heat may be redistributed to other parts of the buildings, or discarded.

For fluorescent lighting approximately 75 % of the heat output is by convection, whereas for tungsten lighting the figure is 60 %, because the lamps run at higher temperatures and hence have higher radiant output. Thus air extract over fittings for internal temperature control is generally more effective for fluorescent lighting.

Local extract above pieces of equipment giving off heat is also employed and is often combined with the removal of unwanted airborne contaminants (Section 4.2). Extract hoods in kitchens are common examples, and numerous other examples occur in industry. If the ventilated air is contaminated, it must be cleaned before any recycling is possible.

Alternatively heat may be extracted from the contaminated air and transferred to incoming fresh air (Section 4.9.4).

Control of gains from solar heating and high air temperatures can be achieved through appropriate design considerations of form and fabric, but if these are not adequate then cooling plant must be employed. Further discussion on these points occurs in Chapters 4 and 5.

3.5.4 Removal of Heat Gains

If gains cannot be controlled at source then they could be removed by ventilation. In the simplest case the ventilation rates are increased, but if the external air temperature is itself high then this technique is limited, although the increase in air movement may assist in comfort (see Section 2.2.3).

If ventilation rates are limited or external air temperatures are high, then ventilation removal can only work if the air is cooled before entry into the habitable spaces. Simple calculations of the ventilation rates for these circumstances follow.

3.5.5 Simple Calculation of Cooling by Ventilation

A small office print room is considered, of volume 120 m^3.

Casual gains. During normal working the following casual gains occur:

3 people at approximately 150 W each (light work)	450 W
Copying and print equipment (obtained from rating of equipment)	2 250 W
Total lighting, assumed to be in use, 6 × 80 W tubes	480 W
Total	3 180 W

External gains. Only the instantaneous heating through glazing is taken into account.

The print room has 2 m^2 of vertical glazing facing south. The latitude is 50 °N. Consider conditions of solar gain during August when external air temperatures are relatively high. The maximum solar intensity on a vertical surface facing south is 530 W/m^2 (see Table 3.14). Therefore the solar intensity falling onto glass = 2 × 530 = 1060 W. The windows are single glazed and clear glass. Therefore the solar gain factor is 0·77 (Table 3.16).

Hence the solar gain through the glass = 1060 × 0·77 = 816 W.

Total internal gains are 3996 W

Ventilation. The printing equipment incorporates localised extract ventilation to remove contaminants. The extract rate is $0{\cdot}4\,m^3/s$. An input fan on the outside wall is arranged to match this flow rate. $0{\cdot}4\,m^3/s$ corresponds to $1440\,m^3/h$, which with a room volume of $120\,m^3$ gives a ventilation rate of 1440/120, i.e. 12 per hour.

The heat transfer by ventilation is given by (Section 3.3.4)

$$Q_v = 0{\cdot}36 \times \text{RVH} \times V \times \Delta t_a$$

If the room is maintained at an air temperature of 22 °C, and the total heat gain in the room is removed by ventilation, what outside temperature is required to achieve this?

Rearranging the ventilation heat transfer equation gives

$$\Delta t_a = \frac{Q_v}{0{\cdot}36 \times \text{RVH} \times V}$$

Substituting values

$$\Delta t_a = \frac{3996}{0{\cdot}36 \times 12 \times 120}$$
$$= 7{\cdot}7\,°\text{C, i.e.} \approx 8\,°\text{C}$$

Hence if the outside temperature is approximately (22 − 8) °C, i.e. 14 °C, then a temperature of 22 °C can be maintained in the print room by direct ventilation. If the outside temperature falls below this, some additional heating of the air into the room will be required to maintain the room temperature.

Consider the outside conditions rising to an air temperature of 18 °C. Δt_a is now 4 °C, i.e. half the previous value. Hence to maintain 22 °C in the space the ventilation rate must be doubled. This may be possible by opening windows or doors, or installing boost mechanical ventilation. But if the ventilation rate cannot be increased, for instance in still air external conditions, then the internal temperature will rise (see Chapter 4 for discussion of practical means of achieving ventilation).

If the outside temperature is 25 °C and the ventilation rate remains at 12 per hour, then, by this simple analysis, the inside temperature will rise to $25 + 8 = 33$ °C which could be considered as intolerable for working conditions. Boosted ventilation would lower this to a certain extent. However, under these circumstances the only real solution is to offer positive temperature control through the use of cooling systems.

Such a system will cool outside air to the required temperature difference in this case of 8 °C (an appropriate value for mechanical cooling; see Section 5.3.5). Hence, input air is still fed into the room at 14 °C to maintain the air temperature in the space of 22 °C. Control of temperatures is also maintained when inside and outside conditions vary while the ventilation rate is kept constant.

Such systems and design considerations relating to their application are discussed in Section 5.3.

The analyses have been performed in a very simple fashion. No account has been taken of fabric heat gains or losses, of the water vapour content of the air (which would affect latent heat transfer) or of solar control. However, the calculations have demonstrated some of the possibilities and limitations to controlling heat gains by ventilation.

3.6 THERMAL INSULATION MATERIALS

3.6.1 Introduction

The steady state approach to heat transfer in buildings has demonstrated that energy losses can be reduced either by controlling heat losses through the fabric or through ventilation. Constraints on minimum ventilation rates have been discussed in Section 2.4, and Chapter 4 discusses practical ventilation control. Fabric heat transfer, given a particular building envelope, can be reduced by lowering the *U*-value of external fabric, which is achieved by incorporating materials of high insulation value into the construction.

The principle behind all insulating materials is to trap small pockets of air (or sometimes another gas) within the material. This is because air has a very low thermal conductivity value ($k = 0{\cdot}025$ W/m °C) compared to solid material—for instance, solid plastic ($k \approx 0{\cdot}2$ W/m °C). The air must be in small pockets to avoid any localised convection heat transfer effects. Thus the best insulation materials are those with a very high proportion of air or other suitable gas in the material, sometimes up to 99 %, i.e. materials with very low densities, less than 50 kg/m^3. Resulting thermal conductivity values for trapped air materials are approximately 0·035 W/m °C.

Because such materials are essentially voids very loosely linked together, they are inherently mechanically weak; many insulating materials are unable to support themselves. If usable mechanical strength is required, then more solid material must be used, and the insulation value naturally decreases. Thus there is an inherent incompatibility between structural and thermal insulation properties.

3.6.2 Types of Insulation Materials

There is a very wide range of insulation materials, and for any one material there is usually a large number of product types. However, the materials can conveniently be grouped into four categories based on the original material source. These are as follows:

Natural organic. The most common in this category are wood based materials, e.g. fibre board or 'insulating' board. Woodwool slab, widely used structurally for roofs, has useful insulation qualities. There is also a wide range of board materials based on other plant products, e.g. from cane, straw, flax and cork.

Artificial organic. Any man made plastic material which is in a 'cellular' or 'foamed' form behaves as an insulation material. Common types are foamed urea formaldehyde (used in insulating cavities by injection), expanded polystyrene, foamed polyurethane (rigid or flexible) and foamed PVC. Some of these materials, such as rigid polyurethane, have gases other than air (e.g. carbon dioxide) trapped inside the cells, which lower the k-value. The types available are summarised in BRE Digest no. 224 [3.23].

Mineral fibres. Fibres of glass or certain natural rocks are very widely used insulation materials because of their low cost and long term durability combined with very good insulation values. A wide range of quilts and boards of varying densities and compressibilities are available, while the fibres may be injected into cavities in loose form.

Mineral—foamed, expanded or cellular materials. These include exfoliated vermiculite, perlite, foamed cement, pulverised fuel ash (PFA), foamed slags and clay and foamed glass.

Most of these materials have usable mechanical strength, particularly if bonded by cement or plaster to produce, for instance, lightweight blockwork. Insulation values, though, are reduced because of increased densities.

Examples of these insulating materials and their properties are given in Table 3.17 and commercial products are summarised in the Insulation Handbook [3.24].

3.6.3 Design Considerations for the Selection of Insulation Materials

(*a*) *U-value required*. For pure insulating requirements this is the major consideration, provided other aspects (such as fire resistance) are met.

Widely used materials are glass or rock fibre quilts, provided that they can be supported or secured in place. Fifty millimetres of such material incorporated into a non-insulating construction typically lowers the U-value to 0·6 W/m^2 °C. The thickness used (e.g. in a loft) is usually a balance

Table 3.17 Some properties of insulating materials (from [3.6, 3.23])

Material	Density (kg/m^3)	k-value (W/m °C)	Typical applications	Comments
Mineral:				
Glass or rock fibres	25–180	0·035–0·042	Wide range	Quilt, slab or loose forms
Cellular glass	130	0·06	'Inverted roofs'	Impermeable to water and water vapour
Exfoliated vermiculite	80–110	0·06	Loose fill or combined with plaster/concrete	High temperature resistance
Expanded perlite	65	0·04	Loose fill or combined with plaster/concrete	High temperature resistance
Lightweight concrete	400–1 600	0·12–0·6	Insulating walls	Wide variety of types
Natural organic:				
Fibre board	250–340	0·06	Decking for roofs	May be bitumen or oil impregnated for water repellance
Woodwool slab	400–600	≈0·1	Structural decking	Relatively inert
Straw board	340–430	≈0·1	Decking	Susceptible to fungal attack in moist conditions
Cork board	120	0·04	Internal linings	
Cellular plastics:				
Expanded polystyrene	16–40	0·035	Under screed, linings, roofs	High vapour resistivity; fire risks
Foamed urea formaldehyde	8	0·043	Cavity fill	Not self-supporting
Expanded PVC	40–70	0·035–0·043	Sandwich constructions	Useful mechanical strength
Foamed polyurethane (rigid) (containing fluorocarbon or carbon dioxide gas)	32	0·01–0·07	Complex shapes, foamable *in situ*	Limited temperature resistance (100 °C)
Foamed polyurethane (flexible)	30	0·035	Pipe insulation	Limited temperature resistance (100 °C)

between the cost of the material and its installation compared with the fuel savings gained. The relationship between thickness of insulating material and U-value is shown schematically in Fig. 3.2. As the real cost of fuel rises so the optimum economic thickness increases, for given climatic conditions (see, for instance [3.14, 3.25, 3.38]).

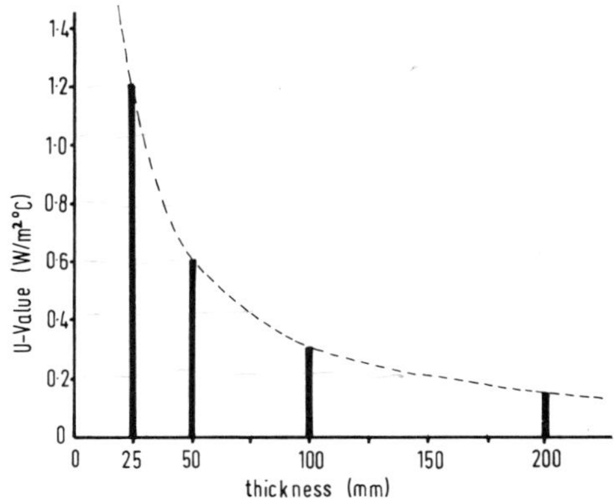

Fig. 3.2. U-value versus thickness for glass fibre insulation. For a doubling of thickness the U-value is halved.

(*b*) *Space available*. If there are limitations on space, for instance with an existing building being internally lined, or with sandwich materials (e.g. glass reinforced plastic) used for cladding panels, then a material with as low a k-value as possible is preferred, so that minimum thickness is required to achieve a given U-value. An example is foamed rigid polyurethane containing trapped fluorinated hydrocarbon with a k-value as low as 0·01 W/m °C. The material can also provide some usable strength in sandwich constructions or board materials.

(*c*) *Structural qualities*. If the structural and insulation qualities of a material must be combined, for economy or ease of construction, then some sacrifices must be made on k-values. Typically, for walls, insulating concrete blocks, either foamed or with lightweight aggregate, are used, and for roofs, woodwool slab or dense fibre board. Lightweight concrete blocks for the inner leaf of a conventional structural brick cavity wall gives an overall U-value of approximately 1·0 W/m^2 °C. For lower U-values it would be uneconomic to increase the thickness of the blockwork because of its relatively low k-value. Hence other insulation material must be incorporated, either in the cavity or else through internal lining.

Insulation materials may be required to have reasonable impact resistance if they are to be used as internal finishes. Again a compromise is usually required, and often the material is given a surface lining, e.g. with plasterboard.

If insulation linings are to be added to an existing roof, for instance a large span factory, then consideration must be given to the extra structural loadings arising from the dead weight of the insulation and internal lining plus the additional snow loads likely due to a colder external roof surface.

(*d*) *Long term stability*. This topic includes resistance to rot and insect attack. It is a particular problem with natural organic materials, especially when they become moist. Protection can be afforded by initial treatment of the material, e.g. with water repellants or anti-fungal agents. Providing the materials are dry, however, they remain stable, and because they have been denatured compared to the original timber or source material there is less risk of attack from beetles and other boring insects. Vermin such as birds and mice can also be destructive to insulating materials.

(*e*) *Water vapour passage*. Because of the open nature of most insulation materials, water vapour passage occurs very readily. Even closed cell plastic materials allow some vapour transfer. This aspect is particularly relevant to interstitial condensation problems, as discussed more fully in Section 3.7.

(*f*) *Water transfer*. Liquid water can soak into many insulating materials and considerably reduces the insulating properties. If this is a likely problem in construction then a closed cell plastic material, e.g. foamed polystyrene, or materials such as rock fibre treated with water repellants, can be used. These do not totally eliminate water penetration, but reduce it to a low level. With such materials, though, the most difficult problem to combat is joints, cracks or fissures across which water can flow. This is particularly relevant with cavity wall insulation [3.26, 3.27].

(*g*) *Stability in fire*. Surface spread of flame, combustibility and fire propagation are properties that must be considered for any building application, and the material chosen must comply with any regulations that apply. Spread of flames within internal linings can be a particular problem as are the toxic fumes given off by plastics. In general inorganic or mineral materials will be most appropriate if fire protection is of importance [3.28].

High temperature stability, e.g. in the insulation of hot water pipes, may also be a criterion.

(*h*) *Acoustic properties*. Internal linings of thermal insulation material can also be incorporated to give a reduction of internal noise levels and reverberation time by providing internal acoustic absorption. This is usually achieved with perforated hard internal surface material, e.g. dense fibreboard, backed by insulation such as glass fibre quilt.

However, when acoustic insulation is required from a wall, i.e. the sound attenuation through the wall should be high, then lightweight construction for thermal insulation becomes incompatible with acoustic requirements. This is illustrated by the more widespread use of lightweight blocks for party walls between dwellings [3.29] because these blocks are already specified, for thermal reasons, for the external walls. The mandatory acoustic attenuation can be achieved from the party wall provided that it is adequately plastered, but improvements would be obtained using heavyweight blocks, when the performance is not so dependent on the plastering process and quality.

(*i*) *Health hazards*. There has been increasing concern over the health hazards associated with insulating materials (e.g. [3.30]). The hazards fall generally into two categories. The first relates to fibrous materials which can cause both skin irritation and more seriously respiratory damage from the inhalation of airborne particles. This is highlighted by asbestos (e.g. [3.31]) but much attention has also recently focused on man made mineral fibres such as those from glass or rock [3.32, 3.33].

The second category of hazard is associated with toxic vapours produced from foamed plastic materials either during foaming processes, or from slow outgasing of materials during their life. Formaldehyde vapour, arising for instance from urea formaldehyde cavity fill foam, has received particular attention (see Section 2.4.7).

Hazards are most likely to occur during installation (or demolition work) and this may influence the specification of material or the manner in which work is undertaken. For asbestos and its derivatives there are well documented recommended health and safety procedures for handling or working materials [3.34]. However, such materials may also present long term exposure risks to occupants of buildings, as outlined in Section 2.4.7.

3.7 CONDENSATION EFFECTS FROM INSULATION

One of the major problems that can arise with increased insulation standards is the enhanced risk of interstitial condensation, i.e. condensation within the fabric of the building. Conditions leading to surface condensation have already been described (Section 1.7.2).

Interstitial condensation arises when water vapour passes from the warm side of the building construction through to colder regions where condensation conditions prevail. This becomes increasingly likely when a wall is internally insulated because of the more steep temperature drop which occurs within the construction.

The effect is well documented (e.g. [3.15, 3.35, 3.36]) and the analysis and prediction of its occurrence are based on examining temperature gradients through the construction, from inside to outside, the driving force for the water vapour flow, namely the vapour pressure difference (see Section 1.7.2), and the permeability of the various components of construction to the passage of water vapour.

The temperature gradients through a construction are determined by proportioning the total temperature drop from inside to outside with the thermal resistances of the materials involved. An insulating material will produce a steep temperature drop across it as it provides a high resistance to heat flow. This is best illustrated by looking at temperature gradients across a solid brick wall, with and without insulation on the inner face, as shown in Fig. 3.3.

The simple wall has an internal surface temperature of 12·5 °C, the drop

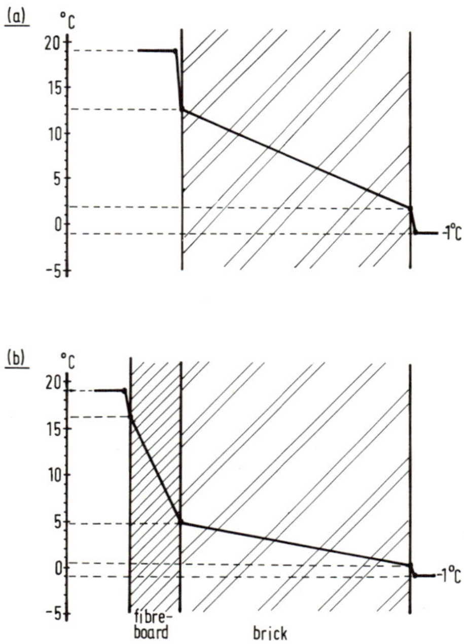

Fig. 3.3. Temperature gradients (from 19 °C inside to −1 °C outside) through simple constructions: (a) a solid masonry wall (220 mm) and (b) a solid masonry wall lined with fibreboard insulation (50 mm) on the inside.

compared to the internal air temperature being due to the surface resistance of the wall.

The addition of insulation to the inner face causes the surface temperature of the brick to drop to 5 °C, which is likely to be below the condensation temperature of water vapour in the room. In practice this would mean that, in typical winter conditions, water would be running down the wall behind the insulation.

The passage of water vapour through a material, given a vapour pressure difference across the material, depends on the vapour resistivity of the material. This is a measure of the difficulty with which water vapour passes through the material and is analogous to thermal resistivity. Typical values are given in Table 3.18, together with vapour resistance values derived from the material thickness (in metres) divided by the resistivity (cf. thermal resistance).

Some materials provide very little resistance to vapour flow, e.g. fibre

Table 3.18 Vapour resistivity and resistance for stated thickness of common building materials (from [3.35])

	Vapour resistivity of materials for unit thickness (MN s/g m)	Vapour resistance for stated thickness	
		Resistance (MN s/g)	Thickness (mm)
Foamed urea formaldehyde	20–30	0·5–0·7	25
Woodwool slab	15–40	0·7–2·0	50
Fibre building board	15–60	0·2–0·7	13
Brickwork	25–100	2·5–10·0	100
Concretes	30–100	3·0–10·0	100
Plasterboard	45–60	0·4–0·6	9·5
Timbers	45–75	2·2–3·7	50
Compressed straw slabs	45–75	2·2–3·7	50
Plaster	60	0·75	12
Rendering	100	1·25	12
Foamed polyurethane (open or closed cell)	30–1 000	0·7–25·0	25
Expanded polystyrene	100–600	2·5–15·0	25
Hardboard	450–750	2·3–3·8	6
Plywoods	1 500–6 000	7·0–30·0	6
Expanded ebonite	11 000–60 000	250–1 500	25
Average gloss paint film	$>10^6$	7·5–40	$<0·1$
Polythene sheet	$>10^6$	250	$<0·1$
Aluminium foil	$>10^6$	4 000	$<0·1$

quilt, while others such as polythene sheet allow virtually no vapour transfer, having very high vapour resistivities; these are known as vapour *barriers*. Vapour *checks* give a resistance of typically 10 MNs/g.

The effects are summarised by plotting the temperature through the wall together with the dew point of the moisture passing through the wall, derived from the vapour pressure and temperatures involved (Section 1.7.2). If the dew point line falls below the temperature line then interstitial condensation can occur. The plots shown in Fig. 3.4 are for winter conditions of 19 °C air temperature and 50 % relative humidity inside and −1 °C and 100 % relative humidity outside, although the effects of other conditions can be established by standard calculation [3.35].

If interstitial condensation conditions are likely to arise, and this is invariably the case with constructions with insulation towards the inside, then the usual means of avoidance is by restricting the vapour passage into the construction initially by incorporating a vapour barrier, e.g. polythene

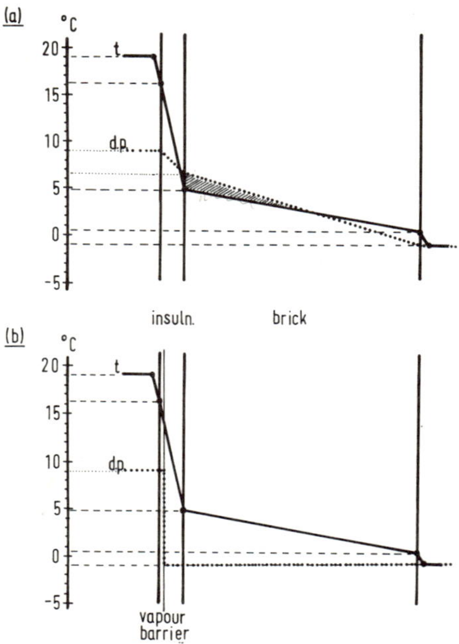

Fig. 3.4. Temperature (t) and dew point (dp) plots through a solid masonry wall with insulation on the inside, (a) without a vapour barrier and (b) with a vapour barrier towards the warm side of the insulation. The hatched area in (a) represents condensation risk. Conditions shown are inside: 19 °C, 50 % RH (dewpoint ≃ 9 °C); outside: −1 °C, 100 % RH (dewpoint = −1 °C).

or aluminium foil, within the construction, as shown in Fig. 3.4(b). This vapour barrier must be positioned on the warm side of the insulation, i.e. the water vapour must be prevented from reaching cold zones. The main problems with vapour barriers are the practical issues associated with fixing without puncturing, overlapping of barriers effectively, and detailing at junctions of walls with floors and roofs.

For constructions where insulation is towards the outside face, where condensation within the fabric will not cause undue problems or where a cavity exists where condensed water can be evaporated away, then interstitial condensation is less of a problem. However in general any wall or roof construction must be examined carefully from the point of view of interstitial condensation risk.

The other major aspect relating to condensation risks from insulation concerns the lowering of ventilation rates and likely increased relative humidity values. This is usually associated with the effective draught proofing of openings, but the addition of insulation linings to a building can reduce considerably the leakage ventilation by sealing cracks in the building fabric, and hence the overall ventilation rate. The likely problems resulting have been discussed in Section 2.4.2.

REFERENCES

3.1 Chartered Institution of Building Services: *CIBS Guide A3: Thermal and Other Properties*. CIBS Publications, London, 1971.

3.2 American Society of Heating, Refrigeration and Air-conditioning Engineers: *1977 Fundamentals Handbook*. ASHRAE, New York, 1977.

3.3 Burberry, P.: *Environment and Services*. B. T. Batsford, London, 2nd edn, 1975.

3.4 Tutt, P. and Alder, D. (Eds): *New AJ Metric Handbook*. Architectural Press, London, 1979.

3.5 Brick Development Association: *Bricks, Their Properties and Uses*. The Construction Press Ltd, Lancaster, 1974.

3.6 Building Research Establishment: *Digest 108: Standard U-values*. HMSO, London, 1975.

3.7 Building Research Establishment: *Digest 145: Heat Losses Through Floors*. HMSO, London, 1972.

3.8 Chartered Institution of Building Services: *CIBS Guide A1: Environmental Criteria for Design*. CIBS Publications, London, 1978.

3.9 Chartered Institution of Building Services: *CIBS Guide A5: Thermal Response of Buildings*. CIBS Publications, London, 1979.

3.10 Chartered Institution of Building Services: *CIBS Guide A2: Weather*. CIBS Publications, London, 1971.

3.11 Chartered Institution of Building Services: *CIBS Guide A4: Air Filtration*. CIBS Publications, London, 1971.
3.12 Institute of Heating and Ventilation Engineers: *Guide Book A, 1970; Section A5, Thermal Response of Buildings*. IHVE (now CIBS), London, 1970.
3.13 Chartered Institution of Building Services: *CIBS Guide B4: Water*. CIBS Publications, London, 1972.
3.14 Burberry, P.: *Building for Energy Conservation*. Architectural Press, London, 1978.
3.15 Markus, T. A. and Morris, E. N.: *Buildings, Climate and Energy*. Pitman, London, 1980.
3.16 Billington, N. S.: Estimation of annual fuel consumption. *J. Inst. Heat. Vent. Eng.*, (Nov. 1966), pp. 253–256.
3.17 Department of Energy: *Fuel Efficiency Booklet 7*. Department of Energy UK, London, 1977.
3.18 *The Journal of the Chartered Institute of Building Services*, London.
3.19 Department of Energy: *Energy Management*. HMSO, London; monthly journal.
3.20 Chartered Institution of Building Services: *CIBS Guide B18: Owning and Operating Costs*. CIBS Publications, London, 1972.
3.21 Chartered Institution of Building Services: *CIBS Guide A6: Solar Data*. CIBS Publication, London, 1971.
3.22 Loudon, A. G.: *Summertime Temperatures in Buildings Without Air Conditioning*. Building Research Station Current Paper, CP 47/68, 1968.
3.23 Building Research Establishment: *Digest 224: Cellular Plastics for Building*. HMSO, London, 1979.
3.24 *Insulation Handbook 1981–82*. Comprint Ltd, Watford, England, 1981.
3.25 Building Research Establishment Working Party: *Energy Conservation: A Study of Energy Consumption in Buildings and Possible Means of Saving Energy in Housing*. Building Research Establishment Current Paper, CP 56/75, 1975.
3.26 Whiteside, D.: *Cavity Insulation of Walls: A Case Study*. Building Research Establishment Current Paper, CP 20/74, 1974.
3.27 Building Research Establishment: *Digest 236: Cavity Insulation*. HMSO, London, 1980.
3.28 Building Research Establishment: *Digest 223: Fire Hazard from Insulating Materials*. HMSO, London, 1980.
3.29 Sewell, E. C. and Scholes, W. E.: *Sound Insulating Performance Between Dwellings Built in the Early 1970s*. Building Research Establishment Current Paper, CP 20/78, 1978.
3.30 Muir, D. C. F.: Health hazards from thermal insulation products. *Ann. Occup. Hyg.*, Vol. 19 (1976), pp. 139–145.
3.31 Michaels, L. and Chissick, S. S. (Eds): *Asbestos: Properties, Applications and Hazards*. Wiley, Chichester, 1979.
3.32 Health and Safety Commission: *Man Made Mineral Fibres Discussion Document*. HMSO, London, 1979.
3.33 US Department of Health Education and Welfare: *Criteria for a Recommended Standard ... Occupational Exposure to Fibrous Glass*. NIOSH, Washington, 1977.

3.34 ASBESTOSIS RESEARCH COUNCIL: *Control and Safety Guides 1–9*. Rochdale, Lancs, 1973/78.

3.35 BUILDING RESEARCH ESTABLISHMENT: *Digest 110: Condensation*. HMSO, London, 1972.

3.36 DEPARTMENT OF THE ENVIRONMENT: *Condensation in Dwellings. Part 1: A Design Guide. Part 2: Remedial Measures*. HMSO, London, 1970/71.

3.37 CHARTERED INSTITUTION OF BUILDING SERVICES: *CIBS Guide A3: Thermal Properties of Building Structures*. CIBS Publications, London, 1980.

3.38 DEPARTMENT OF ENERGY: *Fuel Efficiency Booklet 16, Economic Thickness of Insulation for Existing Industrial Buildings*. Department of Energy UK, London, 1980.

Chapter 4

The Ventilation of Buildings

4.1 INTRODUCTION

Ventilation is the introduction of fresh air and the removal of contaminated air. The rate at which fresh air must be introduced to control the environment is governed by the type of contaminant, its rate of release and the maximum acceptable or permitted concentration (see Section 2.4). Ventilation also causes air movement which affects thermal comfort, as described in Section 2.2.

The design of ventilation systems also involves consideration of the position where the fresh air is introduced, the pattern of air movement in the space, the reduction of spread of contaminants, the dilution of contaminated air and the capture and removal of the contaminated air from the space. Legislation may prevent direct discharge of contaminated air to the atmosphere, in which case the contaminants have to be removed or rendered innocuous before the air is released.

All of these factors affect the selection and detailed design of ventilation systems and have a direct effect on the efficiency, energy consumption and inherent cost of operation. Good design in relation to each factor can minimise the amount of ventilation required, thereby minimising the energy consumed if mechanical ventilation systems are used and also minimising the amount of energy lost when conditioned air is thrown away.

The following sections describe the various methods for achieving ventilation with their principal objectives, features, advantages and disadvantages.

4.2 PRINCIPAL OBJECTIVES

The removal of contaminants from the location where they are produced should always be the first aim of ventilation design. In dwellings this is exemplified by flues and cooker hoods, and in industry this principle can be applied with good effect to many processes.

However, some contaminating activities, such as those from human occupation, are not usually generated in a controlled, localised fashion and so the principle of 'dilution ventilation' must be applied instead.

The means by which ventilation can be achieved also have a bearing on whether localised or general ventilation is applied. Natural means of generating ventilation usually imply a general ventilation approach, although localised ventilation can occur. For instance, the 'natural' thermal forces of an open solid fuel fire produce localised removal of the combustion products. Mechanical ventilation provides more specific control, particularly when considering the removal of contaminants from work processes. These aspects are discussed in the following sections; the objectives described apply to all ventilation systems, whether powered naturally or mechanically.

4.2.1 Localised Extraction

Localised extraction is the most efficient means of capturing an airborne contaminant (pollutant). It operates by capturing the pollutant where it is most dense in the air, which is usually at the point where the pollutant is released into the air.

The pollutant is captured by extracting air from the polluted zone at a velocity which is just sufficient to carry the pollutant with it. For dense particulates (e.g. metallic particles) very high velocities are necessary, but at the other extreme buoyant gases and vapours are very easily controlled by the capturing air stream and so only low velocities are required. To minimise the air quantity handled by extract systems, the design should aim at the minimum velocity for successful capture.

The capturing process should aim to entrain as little unpolluted air as possible. This is achieved by using hoods, canopies, shields and flanges around the extract point to confine the region of capture to the zone where the capture takes place, otherwise much unpolluted air will be entrained (e.g. [4.1, 4.2]). This will reduce the size and capital cost of the ventilation plant and its running costs.

4.2.2 Dilution Ventilation

When localised capture is impractical (e.g. body odours) then the extract ventilation should still be positioned where the density of pollutant is greatest (e.g. the region of largest density of occupation). However, the control of spread of the pollutant and its dilution become important functions of the position and characteristics of the supply of fresh air. As the pollutants are diluted by the incoming air before removal, this is termed dilution ventilation (see Section 2.4).

It should be noted that the space always contains the pollutant during its release and for a while after the release of pollutant has ceased. This has two important implications. Firstly, because the extract system removes diluted pollutant, the system must be larger and handle more air than a local extract system handling the equivalent but concentrated pollutant. Hence, dilution systems potentially convey more energy out of a building in the form of conditioned air than local extract systems. Secondly, the acceptable or permitted concentration of pollutant in occupied spaces may be so low that the use of dilution ventilation is inappropriate and local extraction, or even sealed rooms, may be necessary.

Figure 4.1(a) shows air entering an extract aperture. It is clear that the influence on air movement in the space is small and the capture of polluted air is very localised. Hence, dilution ventilation systems rely upon the polluted air arriving at the extract point by other means (e.g. convection or density effects).

4.2.3 Position of Extract

The positioning of the extract point where the pollutant is most dense requires knowledge of the direction in which the pollutant will travel after release. This motion may be caused by density effects (e.g. water vapour or 'steam' at high level in kitchens and bathrooms), convection currents (e.g. smoky conference rooms), process velocity (e.g. wire brushing, grinding wheels) or by the velocity of the surrounding air, but the general principle is to capture the pollutant with the minimum change of its inherent direction of movement.

For dilution systems, the extract point tends to be placed at high level for convected pollutants (e.g. water vapour in shower rooms) and for pollutants less dense than air (e.g. hydrogen from battery charging); and at low level for pollutants denser than air (e.g. the common industrial degreaser, trichloroethylene). But air movement in the space can be stronger than the natural motion of the pollutant and this can cause general distribution of the pollutant, thereby increasing the area of nuisance and reducing the effectiveness of the dilution ventilation system.

For local extract systems, the capture performance is much better. The positioning of the extract is most important with dense particulates released at high velocities, e.g. grinding, whereas fine dusts, vapours, fumes and gases are more easily captured. However, because the capture area is confined to the immediate vicinity of the extract aperture, it must be located immediately adjacent to the point of release of the pollutant, although hoods, canopies, screens and baffles can be designed to increase slightly the area of capture.

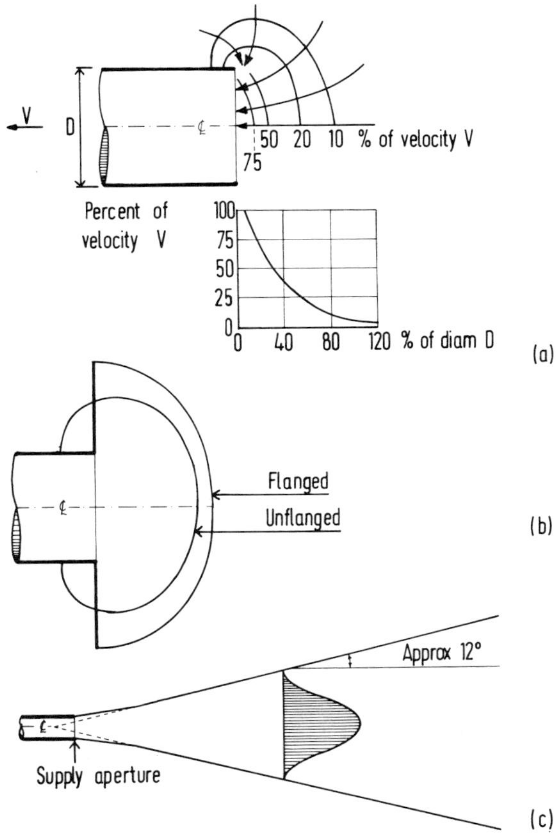

Fig. 4.1. Principal features of air movement at plain circular extract and supply apertures. (a) Velocity contours and air flow paths in the centre line plane of a round extract aperture. (b) The effect of a flange around a round extract aperture on the 10 % velocity contour. (c) A plain free isothermal jet emerging from an opening. The 10 % velocity contour (not shown) occurs at a distance of 30 times the duct diameter from the supply aperture.

Figure 4.1(a) shows air entering an extract aperture evenly from all directions and increasing in velocity as it approaches the peak velocity at the neck of the aperture. The velocity which is necessary to control a pollutant varies from 0·25 m/s for vapours (evaporation) to around 10 m/s for particles released at high velocity (e.g. grinding) [4.1]. It is clear from Fig. 4.1(a) that there are considerable savings in the quantity of air extracted if the control velocity can be achieved by bringing the extract point close to the point of release of the pollutant, otherwise increased air quantities and velocities are necessary to achieve control of the pollutant, resulting in larger systems to both purchase and operate.

The addition of a flange around the extract aperture improves the capture performance and this is illustrated in Fig. 4.1(b). The effect of a flange is to prevent the unnecessary entrainment of air from behind the extract aperture where capture is not usually required. Consequently, flanges concentrate extraction to the zone of capture. The system performance benefits by either achieving the desired capture velocity at a point further from the extract aperture or achieving the same velocity using less extracted air.

The principle of the flange is developed further in the form of shields, canopies and hoods for specific industrial operations [4.3].

4.2.4 Position of Supply

Whereas the effect of the extract point is very localised, the supply point can influence air movement in a space at a considerable distance from the point of introduction of the air if a jet of air is created (see Fig. 4.1(c)). Unfortunately, this feature is rarely of value when capturing pollutants, but it is useful for distributing the fresh air throughout a space and for creating air movement in an otherwise still environment.

The relative position of the supply and extract points can define the general direction of air movement, but of greater importance are the characteristics of the supply. If the supply is in the form of a confined and unobstructed jet, then the penetration into the space will be considerable. The jet will entrain room air and cause additional air movement outside of the region of the air jet. Alternatively, if the supply is evenly diffused and spread over a large area, the influence on air movement in the space will be minimal.

Sometimes, a space requires these two features on different occasions. For example, an office will require a diffused and even introduction of fresh air during cold winter months, with little disturbance of the room air, but during hot summer weather a penetrating jet of fresh air, causing additional air movement, will be appreciated by the occupants due to its cooling effects.

The trajectory and throw of free air jets at the same temperature as the surrounding air can be predicted reasonably accurately when the air discharge velocity and opening dimensions are known, but the prediction becomes increasingly complex when the air jet is of a different temperature to the room air, the jet is projected at angles other than vertical or horizontal, or the room air is moving [4.4]. Further difficulties are introduced if the jet is not free but is projected at or along a surface or junction of surfaces. If many of these complications are encountered, it is difficult to predict confidently the performance of an air jet and the consequential room air movement, although research has produced design methods for some applications [4.5, 4.6, 4.7, 4.8].

As stated previously, the extract position is usually located at the region where the pollutant is most concentrated, whereas the location of the supply of air to the space should not interfere with the removal of the pollutant. The design of the location, velocity, direction and spread of the supply air should aim to achieve the following objectives:

(a) To provide a clean, healthy and comfortable environment for the occupants.
(b) To minimise the spread of pollutant into occupied zones.
(c) Where the pollutants do escape to or are generated in occupied zones, to dilute them in accordance with comfort or health criteria (Section 2.4).

The solutions which stem from the above are individual to each design except for one generalisation:

> The systems should make maximum use of the supply air before exhausting and should avoid the direct passage of fresh supply air to the exhaust point without benefit, namely 'short circuiting' (e.g. [4.1, 4.2, 4.3]).

4.3 METHODS OF VENTILATION

The removal of polluted air and the introduction of fresh air require energy to cause the air to move. This energy can be obtained from natural sources and probably the most common forms of ventilation system rely on the energy in the wind for their operation. The most effective application of wind energy is in 'cross-ventilation'.

Thermal sources can provide energy for inducing air to move vertically through natural convection. The thermal source may be natural, e.g. sunshine heating surfaces, or created by activities such as open fires. The ventilation derived by this method is based on the 'stack effect'.

Finally the energy may be provided by totally artificial means, i.e. using powered fans to create pressure differences which induce the air to move.

There are many variations using this basic theme, but all would be termed 'mechanical ventilation'.

These methods of ventilation are discussed in the following sections.

4.4 CROSS-VENTILATION

When the movement of air is impeded by an obstruction, differences in pressure are caused around the outsides of the obstruction. If the obstruction is a building, these pressure differences can be used to push air into the building and to suck air out when holes (ventilators or openable windows) are placed at positions of increased or decreased pressure (Fig. 4.2).

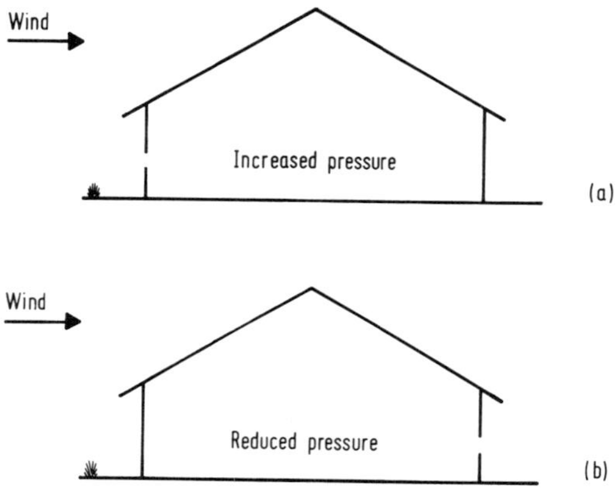

Fig. 4.2. Sections through a building showing pressures generated by air movement. (a) Openings on the windward side. (b) Openings on the lee side.

If one ventilator is in a different room to the other, then a route through the building must be provided for the air to pass from the inlet ventilator to the outlet ventilator. Figure 4.3 shows three plan arrangements in which some of the ventilators could be openable windows.

4.4.1 Applications

The increase in pressure on the windward side of a building is slightly greater than the decrease in pressure on the lee side, and on the other sides and roof the pressure varies in a complex manner [4.9, 4.10].

In approximate terms, the rate of ventilation through a building is

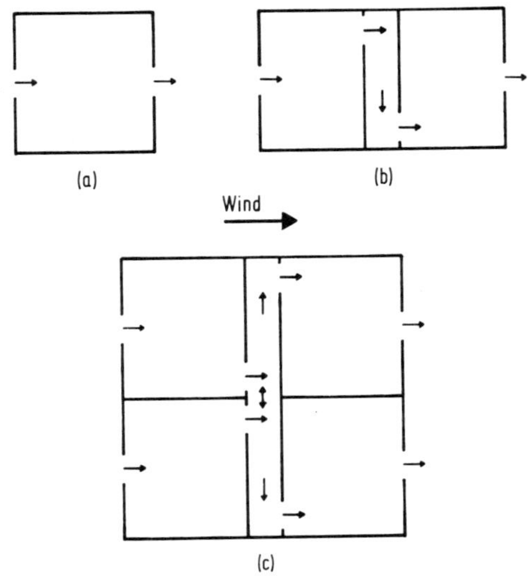

Fig. 4.3. Plan views of air flow through a building by cross-ventilation. (a) Single roomed building. (b) Multiple roomed building. (c) Multiple roomed building with interconnecting air paths.

directly proportional to the wind speed (i.e. double the wind speed and the rate of air flow doubles). Hence, it is possible to design ventilation openings to achieve a particular ventilation rate at a given wind speed and direction, but the design must make provision for the variety of wind speeds and directions which the site experiences.

A simple example will illustrate the magnitude of the ventilation rates involved. A room in a building has two external walls of dimensions 4 m × 2·4 m on opposite sides of the building, which is 6 m deep. Both the external walls contain sliding windows of dimensions 1 m × 1 m which when fully opened provide an opening of 0·5 m^2 on each face. A wind of speed 3 m/s (i.e. a gentle breeze) is blowing at right angles to one face.

There is a simple expression for the ventilation rate when the area of ventilation air inlet is equal to the area of outlet [4.26]:

$$\text{Ventilation rate } (m^3/s) = \text{Constant} \times \text{Open area } (m^2) \times \text{Wind velocity } (m/s)$$

where the constant has a value less than 1 and depends on the geometry of the opening and the wind direction. For this example its value is approximately 0·6. Hence the ventilation rate is

$$0{\cdot}6 \times 0{\cdot}5 \times 3 = 0{\cdot}9\ m^3/s$$

or approximately 56 room volumes per hour. This rate is very high if the removal of typical contaminants is the only criterion involved (Section 2.4.3). However, it is an appropriate value for the provision of adequate air movement during warm weather (Section 2.2).

If the open areas are halved, then the ventilation rate is halved; if the wind speed doubles, then the ventilation rate doubles. Thus cross-ventilation can create high ventilation rates provided there is external air movement and there are openings of reasonable areas on opposite sides of the building.

If doors and windows are shut, the wind pressures will still act on any cracks around door and window frames and on gaps in the construction. Typically ventilation rates of 1 to 2 room volumes per hour will occur in dwellings by this mechanism; these can be reduced to around 0·5 room volumes per hour by sealing the cracks around all openings ('weatherstripping') [4.2, 4.11]. It is important to consider the ventilation which occurs uncontrollably as this may constitute a loss of conditioned air from a building in addition to the designed ventilation rate and therefore more energy may be required to compensate for this loss.

Achieving a particular ventilation rate is simplest in examples (a) and (b) in Fig. 4.3, whereas example (c) is more difficult because the air flow through each room is dependent on the pressure in the adjoining rooms.

The pressures which exist in the rooms along the ventilation air path are dependent on their location, the resistance to air flow at each ventilator and any interconnection with other ventilation air paths (e.g. corridors with rooms on both sides). The prediction of these conditions is cumbersome (e.g. [4.2]) and is not often attempted because the ventilation design may be influenced by other factors, e.g. previous experience, aesthetics and uniformity of design.

The design of the ventilation facilities should provide means for varying the resistance to air flow so that the intended ventilation rate for each space can be obtained under varying wind speeds and directions. But, obviously, it must be accepted that the intended rate of ventilation cannot be achieved by cross-ventilation when there is no wind.

Considering buildings (a) and (b) in Fig. 4.3, the intended ventilation rate is only achieved when the openings are adjusted to give the required resistance to air flow. In theory, opening just one ventilator will not achieve the intended ventilation rate, but if one ventilator is fully open the other may be used as the only control. The maximum benefit from cross-ventilation is obtained when all ventilators remain open continuously, and examples can be found in traditional housing in the tropics where wall constructions contain permanent openings (e.g. [4.12]).

Considering the building (c) in Fig. 4.3, the control is slightly more complicated because there is more than one route for air to flow through the building. Hence, adjustment of one of the openings can also affect the ventilation achieved by other routes. Assuming that the appropriate controllable ventilation openings are provided, then the success of this method of ventilation appears to rely upon the occupants understanding the principles of the system. However, it is doubtful whether the operation of cross-ventilation is generally understood and so success is probably gained through experience and frequent adjustment of the ventilation controls.

4.5 IMPLICATIONS OF CROSS-VENTILATION

4.5.1 Supply Requirements

On the windward side of the building the incoming air jet must be located so that it does not cause discomfort through draughts or differences in temperature, or cause a nuisance by disturbing papers or dust. Frequently, openings are placed above the occupied zone so that in winter a cold incoming air jet has an opportunity to entrain the warmer air in the space when it falls to the occupied zone, and any penetrating air jet cannot cause discomfort immediately.

In summer, large low level openings are often welcomed by occupants because of the penetrating air jets and the increased air movement created, although sometimes the air velocities cause annoyance by disturbing papers. Consequently, ventilation openings are often provided at high level for winter requirements and at low level for summer requirements, often by the appropriate location and sizing of window openings.

4.5.2 Some Supply Options

Some common forms of ventilators in the form of openable windows are shown in Fig. 4.4. One of the more effective ventilation provisions is adjustable horizontal narrow slots, as provided by vertical sliding windows (Fig. 4.4(a)), whereby a very narrow slot can be opened at high level during windy conditions. The incoming air jet quickly disperses its kinetic energy by entraining surrounding room air and when the incoming air is also cold, any problems with cold draughts are minimised by the mixing which occurs with the room air.

During summer, the lower slider can be opened to cause additional air movement in the occupied zone. On still days, both top and bottom sliders

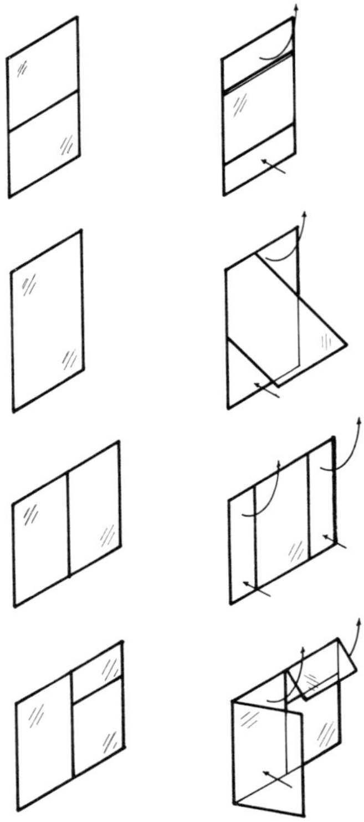

Fig. 4.4. A selection of window designs viewed from outside illustrating typical air flow patterns by the stack effect in winter when the windows are fully open. (a) Vertical sliding window. (b) Centre hung pivot window. (c) Horizontal sliding window. (d) Traditional casement window.

can be opened to use the stack effect (see Section 4.6). Centre hung pivot windows (Fig. 4.4(b)) do not provide such flexibility, sometimes allowing the ingress of cold air at low level in winter when a single opening at high level would have been more successful.

Horizontal sliding windows (Fig. 4.4(c)) allow narrow cracks for windy conditions but again allow the ingress of cold air at low level. The traditional variety of top and side hung casement windows (Fig. 4.4(d)) can produce good results when small opening windows are provided at high level.

Experience gained from the measurement of ventilation rates in a variety of dwellings [4.13] has shown that the usual ventilation controls are much too coarse for the majority of applications. The requirement in winter is for small openable areas at high level with plenty of adjustment from fully sealed to fully open and incorporating the option of a very fine crack opening. This is sometimes achieved through 'trickle' ventilation openings incorporated in window frames. During summer, the coarser control of large openable areas is acceptable.

4.5.3 Room to Room Air Transfer

When the ventilation air path traverses an internal partition, it is usual to provide an air transfer grille. This type of ventilator does not need to be adjustable. Because of space limitations, its size is frequently restricted to that for the winter ventilation rates and during summer the extra ventilation is often achieved by leaving open the internal doors. However, this should not occur with smoke and fire doors as they are usually designed to close automatically after use. The provision of ventilators in doors or partitions also allows more noise transfer and this may exceed the generally expected standard of noise insulation.

If the air transfer grille is mounted in a fire compartment wall, the fire performance of the ventilator must equal that of the wall. Hence, a fire damper of sufficient fire resistance must be incorporated to seal the ventilator when heat is sensed. Fire dampers are available in unpowered forms (e.g. honeycomb matrix coated with intumescent paint or spring loaded steel shutters released by a heat sensitive link) and powered forms (steel shutters operated by springs and solenoids). If it is also necessary to prevent smoke transfer, then powered versions can be linked to smoke detectors (e.g. [4.14]).

4.5.4 Success of Cross-Ventilation

In addition to the problems caused by varying wind velocities there is a further problem to consider. On the lee side of the building, the air being extracted will contain the pollutants collected by the air *en route* through the building. Even though the lee side occupants may open their ventilators wider, the increase in ventilation rate may still be insufficient to reduce the concentration of pollutants sufficiently if the ventilators on the windward side are not open wide enough.

Nevertheless, natural ventilation is still very popular and the generally accepted reasons for its popularity are principally that people can exert individual control over ventilation rates, they usually have sufficient experience with the systems to achieve reasonable success, high ventilation rates can be obtained (especially desirable in warm weather), there is no reliance on mechanical or powered systems and the capital costs are low.

4.5.5 Wind Direction and Location of Ventilators

The design of ventilation systems must provide for the variety of wind directions peculiar to the local climate and site otherwise adverse conditions can cause a design to fail. For example, in kitchens, the atmospheric pollutants (grease, water vapour and cooking smells) are best controlled by powered mechanical extract systems in order to prevent their uncontrolled passage by cross-ventilation through the interior of a building when the wind direction is unfavourable.

Whilst the prevailing wind direction is frequently discussed during design, a study of the meteorological data for a site can indicate the likely frequency of occasions when the wind blows from a different direction. To achieve the intended ventilation rate for a variety of wind directions, it is usual to place ventilators on all external walls when a design permits, thereby achieving cross-ventilation irrespective of wind direction.

Sometimes design restrictions preclude such facilities whereupon ventilators are either placed in one pair of opposite walls, adjacent walls or even just one wall, but with reduced performance. In partial compensation, the size of the ventilator is usually increased but the design relies less upon a calculated solution and more upon the designer's experience. Obviously the principles of cross-ventilation do not apply when ventilators are placed in one wall only, but ventilation may still be achieved with some success because of wind turbulence at the building facade, slight variations in pressure across the facade particularly close to corners or projections, and the stack effect (see Section 4.6.1).

In long rooms with frequent and generous openings in one wall, the effectiveness of summer time ventilation is limited to a depth of approximately 5 m from the external wall, and even less if air movement in the space is impeded by obstructions such as filing cabinets and partitions.

However, such ventilation problems are complex, and can be further complicated by other buildings or obstructions upwind (which sometimes are positioned deliberately to increase sheltering to reduce winter heat loss) and so critical cases require careful examination and development by wind tunnel tests on scale models.

4.5.6 Noise Transfer

Any hole or gap in a wall construction allows airborne sound to pass more readily from one side of the wall to the other. If these openings are in external walls for ventilation purposes then problems can arise through the transfer of noise through the wall, for instance if the building is situated in a noisy environment such as by a major road. Similarly, interference from one room to another can occur or noise can pass out of a building to a quiet residential area. The extent of the noise transfer for a given wall construction depends on the size of the openings involved (e.g. [4.15]). Hence, during summertime, when large ventilation apertures may be open, the problems caused by noise transfer are increased.

In order to improve acoustical attenuation, it is necessary to produce a tortuous air path through the ventilator and to line the air ducts, particularly at the changes in direction, with acoustically absorbent material. Although this reduces noise transfer it also impedes air flow and so it is necessary to increase the size of the ventilators in compensation, which in turn partially increases noise transfer again.

The solution to this problem on a localised basis is to force air through the acoustically attenuated ventilator by using a fan, although this introduces another noise source to be attenuated. For maximum noise control, walls must not be punctured by acoustically weak apertures and so it may be necessary to incorporate mechanical ventilation where the air intake and extract points are sited further away from the noise source (e.g. on the roof of the building) and the system is acoustically designed to minimise noise generation and transfer (e.g. [4.16]).

4.5.7 Ingress of Pollutants in Fresh Air

Where dust, vapours or gases pollute the outside air to an unacceptable level (Section 2.4.6), the introduction of unfiltered air from outside will be undesirable. This can apply equally to an office situated adjacent to a dusty road or railway line as to a hospital operating theatre surrounded by parkland.

Because the usual filtration processes cannot accept the reversal of flow that may occur with natural ventilation systems, the requirement for filtration relies upon powered systems which are also more suited to the relatively large resistance to air flow through the filtration systems (see Section 4.9).

4.5.8 Ventilation Air Temperatures

Sometimes it is not acceptable to introduce unheated ventilation air, such as in very cold climates and buildings for thermally sensitive people or delicate industrial processes. A traditional solution was to site ventilators at low level behind radiators to achieve some heating of the incoming fresh air. This was acceptable when heating systems were operated continuously, but nowadays such designs risk frost damage when heating systems are operated intermittently, and reversed air flow due to a change in the wind direction causes unnecessary energy wastage by extracting heated air from behind the radiator. Manual or automatic dampers are not regarded as an improvement to the design.

A better solution is to introduce fresh air through a system powered by a fan and containing a heater. The system frequently includes filtration to fairly low standards simply to prevent excessive loss of the system's efficiency due to the effect of the build-up of contaminants on the system components, although higher standards of filtration can be provided if required. In addition, acoustical attenuation can be incorporated.

4.6 VENTILATION BY THE STACK EFFECT

4.6.1 Principle of the Stack Effect

The density of air decreases with increasing temperature (Section 1.4). If a vertical column of air at one temperature is connected with a similar column of air at a different temperature, the greater pressure at the bottom of the cooler column will cause air to flow into the warmer column. Taller columns and greater differences in temperature produce greater pressures. This principle has been used traditionally for chimneys and ventilation shafts as illustrated in Fig. 4.5.

The amount of energy required to obtain similar air movement by other means is small in comparison to the heat energy required to cause movement by the stack effect. If this heat energy is obtained directly from natural sources (e.g. solar radiation heat gain onto the sides of ventilation

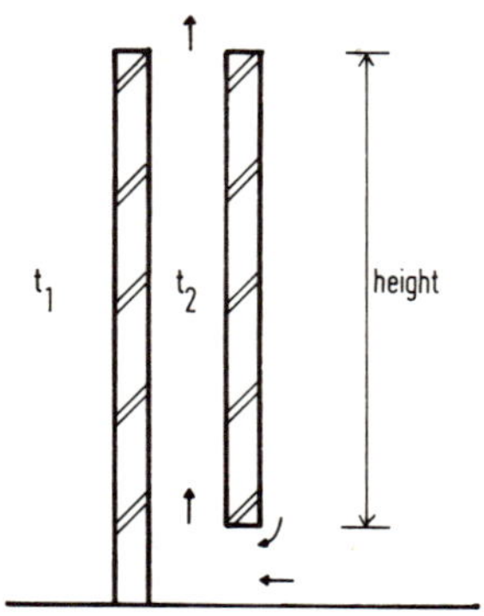

Fig. 4.5. Principle of the stack effect. The effect causes the air flow illustrated when the air temperature t_2 is greater than t_1. If t_2 is less than t_1 air will flow down the stack.

ducts in hot countries) or the purpose is to deliberately dispose of waste heat (e.g. combustion products from incinerators) then the energy losses are of no concern. However, if heated or cooled air is used to achieve the movement then the energy losses can be of great concern.

The difference in pressure between the two columns is proportional to the height of the two columns and approximately proportional to the temperature difference, but the actual air flow is proportional to the square root of the difference in pressure. Hence, a doubling of height or temperature difference causes twice the pressure difference but the air flow rate increases by only about 1·4 (the square root of 2) [4.17].

4.6.2 Applications

The application of the stack effect to the ventilation of rooms requires openings at two levels to allow the air interchange.

During winter, air will enter through the low level ventilator and extract will occur through the high level ventilator. This is useful for removing smoke, water vapour, etc., from occupied rooms. But the reverse will occur during summer if inside air temperatures are cooler than outside air temperatures.

Typically, pressures created by the stack effect are only one tenth of those created by cross-ventilation during a gentle breeze, unless there is a big difference in height between the two openings or a wide temperature differential, such as in the case of chimneys. But even so, a strong breeze can reverse the air flow in a chimney if the wind blows directly onto the discharge aperture. This problem may occur when air movement over the building causes a turbulent air flow directly onto the flue discharge aperture, or the turbulent air flow is caused by upwind obstructions such as other buildings, vegetation and hills.

The effectiveness of ventilation by the stack effect can be seen from an example [4.26]. Consider a cellular office (3 m × 3 m × 2·4 m high) with a vertical sliding window on a calm summer's day. The window is 2 m high by 1 m wide and when opened fully can provide 0·5 m^2 at high level and 0·5 m^2 at low level. (The remaining central area is blanked off by the window panes.) If the outside air temperature is 20 °C and the inside temperature is 25 °C, this will produce a calculated air flow [4.17] of 0·16 m^3/s which removes approximately 1 kW of heat from the room and represents a ventilation rate of 27 room volumes per hour.

If the window (of area 2 m^2) faces west, 1 kW is less than the amount of solar heat transmission experienced in midsummer in temperate zones (e.g. latitude 50° N, Section 3.5.2) and so, because the rate of heat removal is less than the rate of heat gain, the temperature will rise. The air velocity adjacent to the openings is approximately 0·3 m/s, with lower velocities in the room. It should be noted that these conditions are close to thermal comfort for sedentary occupants (Section 2.3) but overheating is imminent.

Comfort would not be experienced in the sun's rays, and any other heat producing equipment will cause more rapid and frequent overheating (e.g. lighting and office equipment) (see Section 3.5.5). Similarly, if the external air temperature was 25 °C instead of 20 °C, internal temperatures would be even higher and discomfort through overheating would be experienced immediately. If such external air temperatures occur on a sunny summer's day then additional means of preventing discomfort must be considered (e.g. solar shading or mechanical cooling, e.g. [4.18]).

4.7 IMPLICATIONS OF THE STACK EFFECT

The previous example shows that the stack effect is useful for the ventilation of rooms with ventilators on one side only but it also illustrates that large openings are required when the vertical distance between ventilators is small, and the system cannot prevent discomfort throughout the summer in moderately warm climates.

The extent and frequency of discomfort in the example could be reduced by:

(a) Reduction of solar gain by, for example, external shading.
(b) Increasing the ability of the structure to absorb heat by, for example, the use of bare concrete or brick (Section 5.2).
(c) Sealing the external facade and reducing the internal temperature using a mechanical cooling and ventilation system (Section 5.3.5).

Simple mechanical ventilation has not been listed above because it would not be of great benefit. The plant would have to be relatively large to supply and extract air quantities in excess of 0·16 m^3/s (27 room volumes per hour) in the example quoted, although it could distribute the air deeper into the space than the 3 m achieved by the stack effect from the single window and also distribute air into rooms without external walls. However, the capital cost of a large ventilation system is often similar to the capital cost of a small mechanical cooling unit and so cooling systems are often preferred because temperatures can then be controlled more accurately with resulting benefits to comfort.

When using cooling systems, ventilation is only required for respiration, odour and pollutant control. The overheating problem is resolved by consuming energy to remove the excess heat from the room.

Occasionally, by preventing solar radiation from actually reaching the window (external shading), approximately 90% of the solar heat gain can be prevented without consumption of energy. The zone adjacent to the window can then be occupied. Nevertheless, cooling systems may still be required to remove the heat from internal sources, the heat introduced by the ventilation required for the occupants and for pollution control (see Table 2.4). This ventilation could be achieved by cross-ventilation or the stack effect, but it is more usual in such cases of heat gain to provide mechanical ventilation so that the amount of ventilation does not vary with wind effects and temperatures (Section 4.9).

4.7.1 Noise Transfer, Pollutants and Temperature of Fresh Air

Because the stack effect usually causes much smaller differences in pressure than those created by cross-ventilation, it requires relatively large openings and its application is restricted by the same inability as cross-ventilation to deal with noise attenuation, filtration and the heating of the incoming fresh air (see Sections 4.5.6, 4.5.7 and 4.5.8).

4.8 COMBINATION OF CROSS-VENTILATION AND THE STACK EFFECT

In Section 4.4.1 it was concluded that one of the problems with cross-ventilation was that little ventilation would occur in almost windless conditions because of the small differences in pressure created across a building. In Section 4.7 it was concluded that one of the problems with ventilation by the stack effect was that of obtaining a significant difference in height between the two ventilators.

Improved ventilation could be obtained by combining these two forms of ventilation, provided the pressures created by cross-ventilation are complementary to those created by the stack effect.

This can be achieved by creating a stack terminated by an extract ventilator projected above the building and directed to discharge with the wind. When wind directions vary, the ventilator can be designed to rotate. Normally, the wind acts upon a vane to turn the ventilator to discharge down wind. Hence, the suction effect across the ventilator and the positive pressure on the windward side of the building provide the cross-ventilation, and the difference in height between the inlet and outlet provide the stack effect. Examples of this combination are oast-house ventilation and the ventilation of houses in hot countries (e.g. [4.19]).

4.9 MECHANICAL VENTILATION

4.9.1 Introduction

Mechanical ventilation is simply the introduction of fresh air and the discharge of polluted air using powered systems (fans) to cause the air to move. The operation of powered systems overcomes problems caused by the capricious nature of wind speed, wind direction and air temperature, and so the performance is more reliable than that from cross-ventilation or the stack effect. However, whenever this degree of reliability is not essential, generally it is preferred to avoid the use of mechanical ventilation because of initial cost, noise, maintenance and the reliance on a power supply. Also the purpose of a mechanical ventilation system (particularly localised systems) may not be understood by the occupants and so the full benefits are frequently not realised [4.27].

The simplest mechanical ventilation systems consist of extract only systems, often positioned to remove specific contaminants. The replacement fresh air enters the building through cracks and gaps in the construction.

The design relies upon the reduction of pressure within the building drawing air through any available aperture, and the ventilation systems may fail to achieve the intended ventilation rate when installed in well sealed buildings.

Mechanically reducing the pressure within buildings can cause other problems, e.g. draughts from cracks around windows and doors, dirt staining around apertures, and chimneys failing to work because of opposition to the stack effect and even damage to the structure in extreme cases. So it is appropriate to make provision for air to enter the building as well as provision for the removal of polluted air, particularly in well sealed modern constructions. These provisions range from simple ventilation apertures to mechanical supply ventilation systems, but whatever the system the objectives stated in Section 4.2 always apply. The various design options are illustrated in Fig. 4.6 and are described below.

4.9.2 Extract Only Systems

The removal of air from buildings reduces the inside pressure and so the greater pressure outside pushes air into the building (Fig. 4.6(a)). The

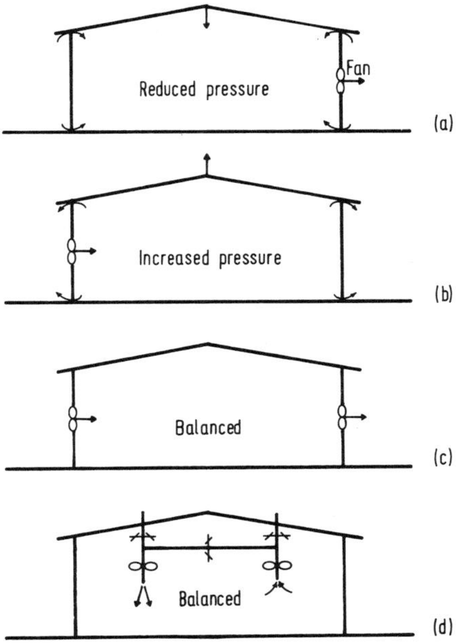

Fig. 4.6. Basic mechanical ventilation systems. (a) Extract only. (b) Supply only. (c) Balanced supply and extract. (d) Central plant arrangement.

difference in pressure created by the extract system will be minimal if there is plenty of opportunity for air to enter the building (e.g. loading bay doors fully open) but the pressure difference will become equal to the maximum capability of the extract system if the building is sealed perfectly whereupon no air flow occurs at all. Hence, before designing the extract system, consideration must be given to the supply air provisions and to the maximum difference in pressure which can be tolerated. (Frequently, extract systems without supply provisions cause difficulty with the opening of entrance doors hinged outwards, and entrance doors hinged inwards swing open violently when released.) The maximum permitted difference in pressure defines the performance of the supply air provisions for the particular ventilation air quantities.

Supply air can be introduced in the required position by a simple hole of adequate size, which is usually covered by wire mesh to prevent entry of birds and vermin. Unless external sheltering is provided, weathered louvres are also required to prevent rain ingress, and the louvres can also be fitted with acoustical absorption materials to provide a small degree of acoustical attenuation. However, additional facilities such as filtration, heating or good acoustical control are usually impractical because of the high resistance to air flow. As this would entail a ventilator of large area in order to prevent excessive pressure differences from occurring, mechanical supply systems are generally used instead.

The extract system could be simply an extract fan mounted at the extract position in the external envelope and discharging through louvres or a cowl. When the best extract position (see Section 4.2.3) is not adjacent to the external envelope it is necessary to duct the polluted extract air to the discharge point and to mount the extract fan at a position in the system where servicing is facilitated.

Statutory control of air pollution occurs in many countries, providing powers to control the discharge of pollutants (e.g. noise, dust and chemicals) which might cause nuisance or harm the environment. Consequently, mechanical extract systems may be necessary to achieve the mandatory degree of control as they can incorporate filtration facilities effectively (e.g. [4.1, 4.3, 4.20]). Noise attenuation may be required to impede noise transfer from inside to outside or vice versa, or to control internal or external disturbance caused by the noise created by the extract system itself (e.g. [4.16]).

The standard of any filtration system will vary with the type of pollutant and the amount of removal required. For example, sawdust from a woodworking factory is not recognised as a general health hazard but the continuous discharge might cause unsightly deposition locally and so filtration of the larger particles would be advantageous, whereas the discharge of some industrial materials (e.g. asbestos fibres and heavy metals) has to be controlled to very fine tolerances. If obnoxious odours in the exhausted air are likely to be a nuisance to others in the vicinity, then it will also be necessary to incorporate deodorising facilities (e.g. [4.21]).

4.9.3 Supply Only Systems

Whereas extract only systems cause depressurisation, it follows that supply only systems cause pressurisation of a building, forcing air out through any available aperture (Fig. 4.6(b)). As discussed previously, it is usual to locate exhaust air ventilators for the best capture of the pollutants, and their size is governed by the maximum amount of pressurisation which can be accepted in the building.

The effects are opposite to those given for exhaust systems only; usually no incoming draughts are experienced at windows and doors, the performance of flues is enhanced and hinged external doors, which open outwards, swing open readily.

Supply only systems are often installed in industrial environments in which light pollutants (e.g. heated air) rise to high level, where they are exhausted by natural means through ventilators in the roof. The supply air system can be designed to direct a jet of fresh air towards occupants who may be close to high temperature heat sources (e.g. furnaces) so that the cool air at high velocity provides more comfortable thermal conditions.

Supply only systems are also applied to the pressurisation of fire escape routes, the objective being to impede the entry of combustion products from the fire into the escape route. The advice of fire control specialists should always be obtained for such applications.

Essentially, the supply air system comprises a fan, weathered louvres or cowl if necessary, wire mesh and sometimes ductwork to direct the supply air to zones where it is required within the building. The system can accommodate noise attenuation, heating, filtration, humidification and other treatment facilities, depending on the requirements in the space (e.g. [4.22, 4.23]).

The natural exhaust ventilators are usually those which can be shut to prevent heat wastage during winter months and overnight. Frequently the ventilators are provided with fusible links so that they open automatically in the vicinity of a fire releasing fumes to the outside. This type of exhaust system offers little noise attenuation and filtration is not usually incorporated.

4.9.4 Combined Supply and Extract Systems

Combined supply and extract systems (Fig. 4.6(c) and (d)) provide the best control of ventilation and offer the opportunity for noise attenuation, filtration and control of the supply air temperature.

The systems are usually designed to slightly pressurise a building thereby opposing draughts through cracks, although strong winds may cause pressures on the windward side of the building in excess of the pressurisation of the building by the ventilation systems, and then air will enter through cracks.

In rooms designed for low levels of airborne dust, pressurisation of the space reduces the uncontrolled ingress of contaminants by air leakage through the fabric and this also assists the arrest of contaminants by double door locks.

Occasionally, when dealing with highly toxic pollutants, the systems are designed to depressurise the polluted rooms or even the complete building to prevent uncontrolled loss of the toxic pollutant.

The temperature control facilities in the supply air (i.e heating or cooling) may be designed to provide ventilation air at a temperature similar to the space temperature so there is no additional heat being supplied to the space or removed from the space to offset other heat losses or gains. Hence, other systems may be required to control temperatures. Alternatively the supply air may provide all the heating or cooling required by the space. These aspects are discussed in Section 5.3.

4.9.5 Energy Recovery with Combined Supply and Extract Systems

Another important advantage of combined supply and extract systems is that there is now an opportunity to exchange energy between the exhaust air and the supply air. Typically, equipment is available to transfer around 75% of the energy which would have been lost, although more energy is required by the fans to push the air through the equipment and the cost of this must be deducted from the value of the energy saved. Nevertheless, such systems are frequently found to be economically viable, often producing a net saving of 60% of the energy which would have been lost.

The variety of energy conservation equipment currently applicable to ventilation systems includes rotating heat exchangers, plate heat exchangers, run-around coils, heat pipes and heat pumps (e.g. [4.24, 4.25] and Section 6.5.4). If the extracted air is polluted to the extent that it is likely to cause reduced performance or blockage of the conservation equipment (e.g. dust and fibres) then filtration of the extract air is also necessary, which increases further the energy consumed by the extract system. All these effects must be considered in the economic analysis of the financial benefits of energy recovery systems (see Section 7.2).

REFERENCES

4.1 CHARTERED INSTITUTION OF BUILDING SERVICES: *CIBS Guide B3: Ventilation and Air Conditioning Systems and Equipment.* CIBS Publications, London, 1972.

4.2 AMERICAN SOCIETY OF HEATING, REFRIGERATION AND AIR-CONDITIONING ENGINEERS: *1977 Fundamentals Handbook.* ASHRAE, New York, 1977.

4.3 AMERICAN CONFERENCE OF GOVERNMENTAL INDUSTRIAL HYGIENISTS, COMMITTEE ON INDUSTRIAL VENTILATION: *Industrial Ventilation. A Manual of Recommended Practice*, 16th edn. ACGIH, Michigan, 1980.

4.4 BATURIN, V. V.: *Fundamentals of Industrial Ventilation.* Pergamon Press, Oxford, 1972.

4.5 JACKMAN, P. J.: *Air Movement in Rooms with Side Wall Mounted Grilles—a Design Procedure.* HVRA Laboratory report no. 65, 1970.

4.6 JACKMAN, P. J.: *Air Movement in Rooms with Sill-mounted Grilles—a Design Procedure*. HVRA Laboratory report no. 71, 1971.
4.7 JACKMAN, P. J.: *Air Movement in Rooms with Ceiling-mounted Diffusers*. HVRA Laboratory report no. 81, 1973.
4.8 HOLMES, M. J. and CAYGILL, A.: *Air Movement in Rooms with Low Supply Air Flow Rates*. HVRA Laboratory report no. 83, 1973.
4.9 BUILDING RESEARCH ESTABLISHMENT: *Digest 119: The Assessment of Wind Loads*. HMSO, London, 1974.
4.10 BUILDING RESEARCH ESTABLISHMENT: *Digest 210: Principles of Natural Ventilation*. HMSO, London, 1978.
4.11 BUILDING RESEARCH ESTABLISHMENT: *Digest 190: Heat Losses from Dwellings*. HMSO, London, 1976.
4.12 KOENIGSBERGER, O. H., INGERSOLL, T. G., MAYHEW, A. and SZOKOLAY, S. V.: *Manual of Tropical Housing and Building. Part I: Climatic Design*. Longman, London, 1974.
4.13 WELLER, J.: *The Computation of the Energy Consumption of Buildings and its Application to the Thermal Performance of Dwellings*. PhD. Thesis, CNNA, Lanchester Polytechnic, Coventry, England, 1977.
4.14 CHARTERED INSTITUTION OF BUILDING SERVICES: *CIBS Guide B5: Fire Protection Systems*. CIBS Publications, London, 1972.
4.15 PARKIN, P. H., HUMPHREYS, H. R. and COWELL, J. R.: *Acoustics, Noise and Buildings*, 4th edn. Faber and Faber Ltd, London, 1979.
4.16 WOODS, R. I. (Ed.): *Noise Control in Mechanical Services*. Sound Attenuators Ltd, Colchester and Sound Research Laboratories Ltd, Sudbury, Suffolk, 1972.
4.17 CHARTERED INSTITUTION OF BUILDING SERVICES: *CIBS Guide A4: Air Infiltration*. CIBS Publications, London, 1971.
4.18 HARKNESS, E. L. and MEHTA, M. L.: *Solar Radiation Control in Buildings*. Applied Science, London, 1978.
4.19 MARKUS, T. A. and MORRIS, E. N.: *Buildings, Climate and Energy*. Pitman, London, 1980.
4.20 CHEREMISINOFF, P. N. and YOUNG, R. A.: *Air Pollution Control and Design Handbook*. Marcel Dekker, Basel, 1977.
4.21 SUMMER, W.: *Odour Pollution of Air—Causes and Control*. Leonard Hill, London, 1971.
4.22 CROOME-GALE, D. J. and ROBERTS, B. M.: *Air Conditioning and Ventilation of Buildings*. Pergamon Press, Oxford, 1975.
4.23 AMERICAN SOCIETY OF HEATING, REFRIGERATION AND AIR-CONDITIONING ENGINEERS: *1975 Equipment Handbook*. ASHRAE, New York, 1975.
4.24 REAY, D. A.: *Industrial Energy Conservation*. Pergamon Press, Oxford, 1977.
4.25 ELECTRICITY COUNCIL: *Heat Recovery for Industry*. Paper EC3793, London, 1978.
4.26 BRITISH STANDARDS INSTITUTION: *Design of buildings: ventilation principles and designing for natural ventilation*. BS 5925, BSI, London, 1980.
4.27 BRITISH STANDARDS INSTITUTION: *Mechanical ventilation and air conditioning in buildings*. BS 5720, BSI, London, 1979.

Chapter 5

Temperature Regulation Systems

5.1 INTRODUCTION

Temperature regulation requires consideration of more than just the control of the temperature of the air in a space. The radiant conditions, the air velocity and the direction of air movement are all important factors. Frequently, the relative humidity must be considered as well, because of its relationship with air temperature (see Section 1.7.2). The most stringent control of temperature is required for some industrial processes, for example colour printing on paper for which both temperature and humidity must be controlled accurately and constantly to achieve dimensional stability and enable high quality reproduction. For the majority of commercial, industrial, educational, social and domestic applications, temperature and humidity control are required for human comfort which, in comparison with the previous example, is relatively lax, although it is probably in this application that most complaints are received!

The least stringent control of temperature is required for a wide variety of applications, e.g. garden sheds, telephone kiosks and warehouses. However, in these examples high relative humidity can sometimes have an undesirable effect on, for instance, the rate of corrosion of metals, the growth of moulds and the strength of cardboard packaging.

So the performance of the temperature regulation system (and in effect humidity regulation) can be more demanding than it might have appeared at first.

Consequently, careful consideration must be given by a building design team to the temperature and humidity requirements in each space so that the building fabric and any services can be designed to achieve the necessary degree of control. Consideration must be given to the range of temperatures permitted, the accuracy of control required, the duration for which the control is required and the consequential relative humidity.

Sometimes sufficient control of temperature can be achieved by the building envelope and natural ventilation systems, and this is called a 'passive' form of control. When better control is required, this usually requires mechanical or electrical services and the inevitable consumption of energy, and so this is called an 'active' form of control. For minimum energy consumption, the passive system should aim to achieve the majority of the control, leaving the active system to do only the final fine control.

The following will examine the manner in which passive and active systems operate thermally, giving detailed consideration to the application of the fundamental heat transfer processes as this helps to improve understanding and so provides the opportunity for reduced consumption of energy by better design.

5.2 PASSIVE METHODS OF TEMPERATURE REGULATION

5.2.1 Damping and Delaying Effects by the Building Envelope

The simplest form of temperature control existed with man's early attempts at buildings to protect himself from the elements. This passive form of temperature control can be recognised today in many forms. The glasshouse offers little resistance to heat loss and solar penetration. The marquee also offers little resistance to heat loss but provides slightly better control of solar radiation. Dwellings frequently offer better resistance to heat loss and sometimes, by careful design of the glazed areas, solar gain is maximised in winter and minimised in summer. In each of these examples of methods of temperature control, the system achieves its own quality of temperature control passively.

To gain a better understanding of passive control, consider a very simple example which will be developed into more comprehensive examples later. In the following examples, any unqualified reference to temperature will imply environmental temperature (see Section 1.6.2) unless specifically stated otherwise (e.g. air temperature, wall surface temperature, etc.).

Consider a single roomed building constructed from solid materials but without any windows, internal heat sources or ventilation. If the external temperature remains constant, the internal temperature will gradually change to the same value as the external temperature. This is because any difference between internal and external temperatures will cause heat to flow through the structure until eventually the temperatures are equalised.

It has been assumed that the external temperature remains constant with time and this condition is called 'steady state'. However, a constant external temperature rarely occurs for any extended period. There are variations due to differences between day and night conditions (diurnal variations) and changing meteorological conditions over short periods (weather) and long periods (seasons). Hence, during building design, it would be most unrealistic to base any detailed thermal decisions on steady state conditions.

Diurnal variations in external temperature influence the internal temperature in the following manner. Starting from steady conditions, consider the effect of a rise in the external temperature on the structure of the simple building. When the external temperature rises, heat will flow into the external surface of the structure by convection and radiation (Section 1.6.1) and so the external surface temperature will increase. Heat will then start flowing into the structure by conduction (Section 1.5). Eventually, heat will arrive at the inside surface and will be noticed as an increase in the inside surface temperature. This is shown in Fig. 5.1(a).

There are two important facets of this transfer of heat. Firstly, there has been a period of time between the instance when the external temperature increased and when the heat first appeared at the internal surface. This is referred to as a time lag.

Secondly, a certain quantity of heat has passed through the structure to

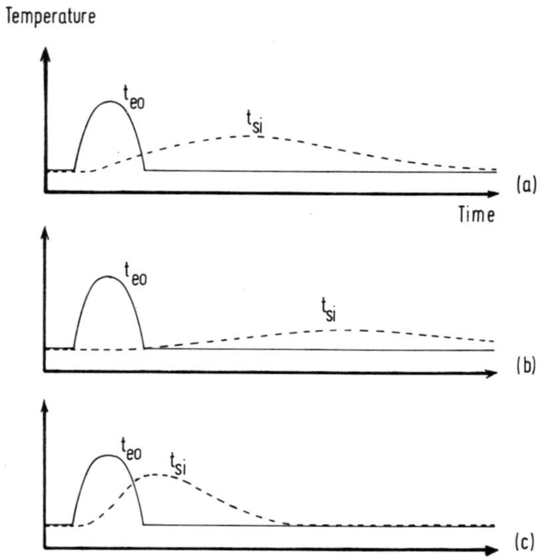

Fig. 5.1. Diagram illustrating the thermal response of a simple building to changing external temperature (t_{eo}: external environmental temperature, t_{si}: inside surface temperature). (a) The effect of a brief rise in external environmental temperature for a medium response windowless building. (b) As (a) but for a heavyweight building with slow thermal response. (c) As (a) but for a lightweight building with a quick thermal response.

cause the internal temperature to vary by a smaller amount than the external temperature. In fact, during diurnal temperature cycles, some of the heat which enters the external surface also leaves by the external surface when the external temperature falls below the external surface temperature. At this stage the temperature within the envelope of this building is higher than both the inside and outside surface temperatures. This feature is referred to as a temperature wave and is a direct result of the fluctuation in the external temperature. The peak of the wave decreases in amplitude with time and travels away from the side on which the cyclic fluctuations occur. So the amount of heat appearing internally is less than the amount which originally entered the external surface. This is related to a term known as the damping factor which is often expressed as the amplitude of the internal surface temperature variation compared to the amplitude of the external surface temperature variation.

The time lag and damping factor are both complex functions of the frequency of the cyclic fluctuations [5.1, 5.2]. During very rapid cycles (e.g. a few minutes of strong sunshine through discontinuous cloud cover) the heat travels through the fabric quickly but it is so severely damped that it is of little significance when it arrives at the internal surface. Conversely, very slow cycles (e.g. weekly or seasonal temperature changes) cause the heat to pass through the fabric slowly but the damping factor is so weak that the influence on the internal temperatures is of much greater significance (see Fig. 5.2).

A practical use of these features is the depth to which uninsulated water pipes are buried in the ground for protection from intermittent freezing weather. Depending on the nature of the subsoil, seasonal variations have little effect on temperatures at depths below about 25 m where subsoil temperatures are close to the annual mean temperature, and diurnal variations have little effect below about 1 m where subsoil temperatures will be close to the weekly mean temperature. Hence, choosing the depth to which water pipes should be buried for protection against intermittent freezing involves examining weather data to determine the time lag necessary and relating this to the thermal properties of the subsoil (e.g. [5.2]).

In addition to the frequency of the cyclic fluctuations, the time lag and damping factor are also complex functions of the thermal properties of the material through which the heat is travelling, namely thermal conductivity and volumetric thermal capacity (Section 1.8). These are combined for convenience and called thermal diffusivity which is the thermal conductivity divided by the volumetric thermal capacity. Typical values are given in

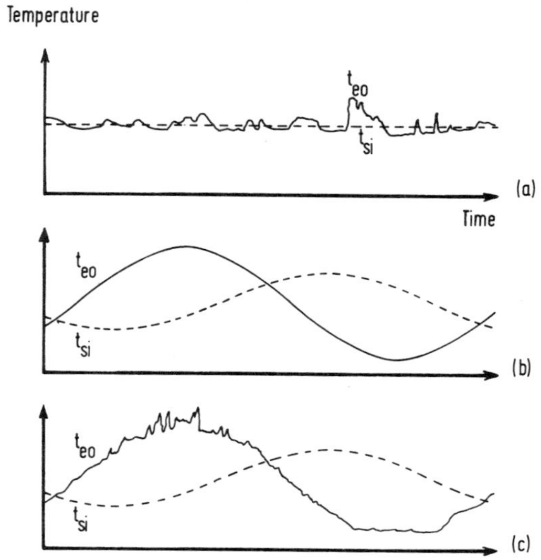

Fig. 5.2. Effect of external environmental temperature (t_{eo}) cycles of different frequencies on the internal surface temperature (t_{si}). (a) Rapid fluctuations of t_{eo}. (b) Slow cyclic fluctuations of t_{eo}. (c) Combined rapid and slow fluctuations of t_{eo}.

Table 5.1. Selecting materials with high diffusivity (e.g. brick, concrete, expanded polystyrene, granite, limestone and slate) causes the time lag to decrease and the damping factor to increase and materials of low diffusivities (e.g. foamed lightweight concrete, cork, plaster, insulating board, slag-wool, wood and chipboard) produce the opposite effect. It is of interest to note that expanded polystyrene, an insulating material, has a high thermal diffusivity because it has a low conductivity and a very low capacity.

Considering the effects of diurnal cyclic fluctuations, Table 5.1 also gives typical velocities of sinusoidal temperature waves for a selection of common building materials. Knowing the thickness of the material, the time taken for a temperature wave to pass through the material during 24 hour cycles can be obtained by dividing the thickness by the velocity. Alternatively, the thickness of the material could be designed to cause a particular delay so that, for example, the peak external temperature causes the temperature wave to arrive at the inside surface at the time it could be most useful.

The amplitude of the temperature wave decreases exponentially as the wave passes through the material and so more damping occurs with

Table 5.1 Thermal properties of a range of common building materials (adapted from Billington [5.1] and Addleson [5.3])

Material	Thermal conductivity k (W/m °C)	Volumetric thermal capacity (density × specific heat) ρc (J/m^3 °C) $\times 10^{-6}$	Thermal diffusivity $k/\rho c$, i.e. κ (m^2/s) $\times 10^6$	Thermal penetration coefficient $k\rho c$ (J^2/m^4 °C^2 s) $\times 10^{-6}$	Velocity of propagation for 24 hour cycle v (m/s) $\times 10^6$	Damping factor t_l/t_0	Thickness l (m)
Brickwork	0·81	1·62	0·50	1·31	8·53	0·43	0·1
Gravel concrete	1·44	2·41	0·60	3·47	9·32	0·64	0·1
Foamed lightweight concrete	0·14	0·67	0·21	0·09	5·51	0·27	0·1
Cork slab	0·04	0·23	0·17	0·01	5·03	0·23	0·1
Expanded polystyrene	0·03	0·02	1·50	0·0006	14·8	0·61	0·1
Glass	1·05	2·11	0·50	2·21	8·5	0·92	0·01
Granite	2·93	2·44	1·20	7·15	13·2	0·58	0·1
Gypsum plasterboard	0·16	0·80	0·20	0·13	5·39	0·86	0·01
Insulating board	0·06	0·41	0·15	0·025	4·61	0·21	0·1
Limestone	1·53	1·92	0·80	2·94	10·8	0·51	0·1
Cement and sand render	0·53	1·31	0·40	0·69	7·67	0·91	0·01
Slate	1·88	2·05	0·92	3·85	11·5	0·53	0·1
Wood	0·14	0·91	0·15	0·13	4·73	0·21	0·1

increasing thickness of material. For 24 hour temperature cycles, Table 5.1 gives typical values for damping factors for a selection of building materials. For the purposes of comparison, the damping factors are quoted for the thicknesses specified in the table and this may not be the actual thickness in a practical application. The velocity of propagation and damping factors for other thicknesses can be calculated from the formulae given in Table 5.2.

From these it is clear that the time taken for a temperature wave to penetrate a material and the amount of damping of the amplitude of the temperature wave both increase with increasing thickness of material

Table 5.1 also gives values of the thermal penetration coefficient which is discussed in Section 5.2.3.

Returning to the simple example, now that the heat has arrived at the inside surface of the walls in the simple building, it will cause the internal

Table 5.2 Expressions for the velocity of propagation and damping factor (from [5.2])

Velocity of propagation (v)

$$v = 2\sqrt{\left(\frac{k\pi}{\rho c\tau}\right)}$$

or

$$v = 2\sqrt{\left(\frac{\kappa\pi}{\tau}\right)}$$

Damping factor (t_l/t_0)

$$\frac{t_l}{t_0} = \exp\left[-l\sqrt{\left(\frac{\rho c\pi}{k\tau}\right)}\right]$$

or

$$\frac{t_l}{t_0} = \exp\left[-l\sqrt{\left(\frac{\pi}{\kappa\tau}\right)}\right]$$

where k = thermal conductivity (W/m °C).
ρc = volumetric thermal capacity (J/m^3 °C).
τ = period of the cyclic fluctuations (s).
κ = thermal diffusivity (m^2/s).
t_0 = the amplitude of the temperature fluctuations at one plane—plane zero (°C).
t_l = the amplitude of the temperature fluctuations at a plane at distance l (m) from plane zero (°C).

temperature to rise in the following manner. The warmed internal surface will cause the air adjacent to that surface to become warmer and to rise by convection to high level (providing the opportunity for natural air movement exists), causing convection currents in the room.

The warm air will distribute itself at high level and cooler air will move onto the surface of the warmed wall at low level to be warmed and rise to high level. This process will continue, forming an increasing depth of warmer, lighter air at high level, whilst the inside surface of the wall is warmer than the air in contact with it. If the wall surface becomes cooler than the adjacent air the process will reverse.

The collection of the warmer air at high level is often not felt by the occupants because the air remains above the occupied zone. However, it does become noticeable and cause over-heating discomfort at higher levels in buildings with occupancy over the full height (e.g. maintenance areas over production equipment or mezzanine floors).

The warmer air, now at high level, may encounter cooler surfaces (e.g. ceilings or partitions) and so some heat will be given up to those surfaces and the air will cool and gradually fall from its high position. Hence, a gentle circulation of air occurs, caused by the surfaces in the room being at different temperatures.

Whilst these convection currents are in operation, a net radiation loss will occur from the warm internal surface of the wall to other cooler surfaces which it can 'see'. So the internal partitions, the floor, the ceiling and the furniture will all gain heat by radiation from the warmer surfaces and therefore become warmer, causing opportunities for more convection currents to arise.

Consider the overall effect of this heat flow on the simple windowless building. The brief rise in external temperature has caused a certain amount of heat to flow into the building interior over an extended period of time and at a time after the external variations occurred, as shown in Fig. 5.1(a). The effect which this quantity of heat will have on the internal temperature will depend on the characteristics of the internal structure of the building and on the building contents. For example, a heavily structured masonry building or a building containing heavy materials in storage, machinery, partitions or floors will absorb the heat with little change in temperature, whereas a lightweight building, devoid of any internal mass, will have little to absorb the heat. Air has little thermal capacity because of the low mass involved (Section 1.8) and so the increase in temperature in the lightweight building will be comparatively large (see Fig. 5.1(b), (c)).

The ability of a building to absorb heat is related to its thermal capacity.

Buildings which are able to absorb large quantities of heat over short periods with little change in temperature are called thermally heavyweight buildings. Buildings which change temperature rapidly when the same quantity of heat enters are called thermally lightweight buildings.

Now consider the case of a temporary fall in the external temperature. The effects are similar to the rising temperature just discussed, except, of course, that the direction of heat flow is reversed. With the fall in outside temperature, heat is lost from the external surface causing a reduced surface temperature—the effect of which eventually arrives on the inside surface as a reduced temperature due to heat conduction from inside to outside. Heat is transferred from the room air onto the now cooler internal wall surface and convection currents gently commence in a downwards direction because the room air is cooled. The cooled, denser air falls to low level and spreads out across the floor forming a gradually deepening layer of cooler air beneath the main bulk of the room air. In occupied buildings, this effect is frequently sensed by occupants because the cooler air enters the occupied zone where it collects at low level causing possible discomfort to the feet, ankles and legs.

Simultaneously, a net radiation loss occurs from the warmer internal surfaces to the cooler wall surface.

The internal temperatures which actually result from fluctuating external conditions can be influenced considerably by the selection of construction materials, the positioning of insulating materials and the type of building contents. For example, a building with a quick thermal response will have its quick thermal response damped if heavy materials are stored within the building, e.g. a warehouse storing steel products. Hence, there would be no benefit in striving to increase the thermal response of such a building by selecting a particular design for the building envelope.

It is clear now that the design of buildings influences the ingress of heat and also influences the effect that that quantity of heat has on the internal thermal environment. Figure 5.1 shows these effects on a thermally heavyweight and a thermally lightweight building respectively.

So far, only a very simple change in the external conditions has been considered but normally the external temperature fluctuates about the mean temperature. Figure 5.2 shows how the internal temperature also fluctuates about the mean external temperature although the internal fluctuations are reduced in amplitude and delayed, according to the frequency of the external temperature cycles and the thermal performance of the building.

Such damping of the daily fluctuations in external temperature by the building envelope can be most beneficial. On occasions, the rise to an unbearably high external temperature during the afternoon is followed by a fall to an uncomfortably low external temperature overnight, whereas the internal temperature can follow quite closely a comfortable mean value if the fluctuations are damped sufficiently. This performance can be achieved without consuming energy.

Many examples of such buildings can be found in the equatorial countries where windows are kept small and the penetration of solar radiation is prevented by external shading features. The envelopes of these buildings provide adequate damping of the external fluctuations in temperature and sufficient time lag so that the afternoon's heat eventually appears internally when temperatures would otherwise have been falling. For the reverse situation during the day, the buildings remain cool when the external temperature is rising.

5.2.2 The Influence of Windows Shaded from Sunshine

In place of the simple windowless building considered in Section 5.2.1 an all-glass building will now be examined. It is assumed that this all-glass building is totally shaded from solar radiation.

As before, the internal temperature will be the same as the external temperature whilst external conditions do not change. If the external temperature now fluctuates diurnally the same effects will occur as in the previous section but with the following important differences.

Firstly, the transfer of heat between the external air and the glass, and then through the glass (which is typically only 5 mm thick) occurs so rapidly that the time lag is of no significance and is usually ignored. Hence, when a change occurs outside, convection and radiation effects begin to appear inside almost instantaneously.

Secondly, because the time lag is insignificant, the heat flow through the glass will simply be proportional to the difference between the outside and inside temperatures at any instant in time. Hence, the damping factor is close to unity (i.e. little damping occurs) and is only a function of the resistance to heat flow between the two glass surfaces and the surroundings.

So the all-glass building responds quickly to outside influences and greater fluctuations will occur internally because of the absence of thickness of material in the envelope. This heat transfer could have a significant effect on the internal temperature. If this all-glass building is thermally heavyweight due to heavy partitions, floors or stored materials, its response will be damped in comparison to an all-glass building which is thermally lightweight, because more heat is required by the structure of the heavyweight building for the same increase in internal temperature.

However, both heavyweight and lightweight buildings will be subject to deviations in temperature which will be much greater than the equivalent windowless building considered previously, and care must be exercised with all-glass buildings which are thermally lightweight, otherwise rapidly fluctuating temperatures may cause problems.

The reaction of a conventional building, containing just a few windows set in the opaque part of the building envelope, can now be obtained by combining the performances of the windowless building and the all-glass building. Following a rise in the outside temperature the glazing will allow an instantaneous heat transfer to inside which will immediately cause internal temperatures to change. Whilst this is occurring, heat from outside will be entering the thicker parts of the building envelope and will gradually travel through to arrive at the inside at some time later, depending on the time lag. Consequently, there are two heat flows occurring simultaneously, which leads to complications when trying to predict the internal temperatures during the course of a change.

The fluctuation in internal temperature due to a change in the external temperature is therefore faster than the windowless building. As more glazing is incorporated so the amount of instantaneous heat transfer increases and the fluctuations in internal temperature become faster and larger until the building is practically all glass and therefore behaves as such.

5.2.3 The Influence of Windows Without Sun Shading

In the previous example no solar radiation was considered to enter through the windows. The following considers the effect of solar radiation penetrating windows and landing on internal surfaces.

Previous discussion has explained how glazing causes a building to respond more rapidly to external influences and how the internal temperatures fluctuate more. To this performance must now be added an input of solar radiation which is penetrating the glazing and landing on a surface somewhere inside the building. The effect will depend on the colour of the surface, the ability of the surface material and substrate to conduct heat away from the surface, and the location where the solar radiation is landing.

Furthermore, the quantity of solar radiation which actually penetrates the glazing is a complex function of latitude, time of year, time of day, weather conditions and atmospheric pollution, combined with the area of glazing irradiated, surface cleanliness and reflection characteristics, orientation with respect to the sun, whether the glazing is vertical, horizontal or angled, the glazing's transmission characteristics and finally how many layers of glass the solar radiation must penetrate (single, double or triple glazing).

Fortunately, solar radiation intensities can usually be obtained approximately from tabulated data or simple calculations as explained in Section 3.5 and [5.4, 5.5, 5.6]. Hence, the following assumes that radiation intensities penetrating the glazing are known.

There are four important implications to consider. Firstly, the solar radiation causes heating within the building instantaneously; secondly, the quantity of heat transmitted through the glazing can be significant (i.e. hundreds of watts per square metre) especially when the sun's rays strike the glazing perpendicularly; thirdly, having passed through the glazing, this heat is then trapped in the manner commonly known as the greenhouse effect (see Section 1.3.6); and finally, some heat is absorbed by the glazing which consequently becomes warmer.

It is now apparent that the glazed building has a thermal performance which is different to either the windowless building or the glazed building shaded from solar radiation. It has an ability to collect solar energy. So whatever temperatures occurred in a building with shaded glazing will be increased by the amount of heat passing through and trapped by the glazing which is not shaded.

When shading is considered to be important, the shading must be designed to prevent the penetration of direct solar radiation. The diffuse radiation scattered by atmospheric constituents (e.g. clouds) and reflected from the ground or other buildings is usually of such a nature that the geometry of the shading is difficult to determine. However, diffuse solar radiation is often much weaker than the direct radiation and so it does not usually cause a problem (see Tables 3.14, 3.15).

The physical characteristics associated with solar radiation penetrating glazing provide designers with the opportunity to influence the time when the instantaneous heat gain occurs. Provided the glazing receives direct solar radiation, the collection of energy in winter will be maximised by facing vertical glazing due south in the northern hemisphere (and vice versa in the southern hemisphere). This arrangement automatically impedes the penetration of the higher summer sun's rays as the glazing reflects more of the oblique radiation and the apparent area of glazing as seen from the sun's position is less than in winter (see Section 1.3.5). Summer penetration can be reduced further by shading the glazing from the sun's rays by the use of, for example, deciduous trees, external blinds and louvres, e.g. [5.7, 5.8, 5.9, 5.10].

Alternatively, the collection of solar energy can be timed to occur in the morning (east facing vertical glazing), the afternoon (west facing vertical glazing) or midday (horizontal glazing or angled towards the noon sun). Such arrangements can be used to influence the fluctuating temperatures which will occur diurnally.

The collection of energy will continue whilst solar radiation is penetrating the glazing. However, if the solar radiation causes the internal temperature to exceed the glazing temperature and the external temperature, there will be a heat loss through the glazing and other parts of the building envelope, and a heat loss by ventilation. Consequently, the internal temperatures will not increase beyond the point when the rate of radiation heat gain through the glazing equals the rate of heat loss through the glazing, plus the heat lost into other surfaces and by ventilation, and the heat conducted into storage in internal partitions, floors, etc.

During the design of new buildings it is advisable to check that this dynamic heat balance produces conditions which will be satisfactory. The study of temperatures under such conditions is an ideal task for a computer which can simulate the conditions and provide the designer with much useful information on temperatures and energy flow during design, so that improvements to the building can be considered and tested (see Section 6.3). However, in order to conduct useful comparisons by computer, the building design team must know how the building should be modified so that improvements are made in the right direction when reselecting building materials to affect the thermal performance. In addition to the properties of thermal conductivity, thermal capacity and thermal diffusivity, consideration must now be given to the properties which influence the manner in which surface temperatures change.

Reconsider the solar radiation penetrating the glazing and landing on surfaces within the room. The relative amounts of absorption and reflection can be influenced by the selection of the surface's absorptivity (see Section 1.3.3) and so the amount of radiation absorbed by the surface can be determined.

If the radiation is absorbed by a thick surface which has the ability to conduct the heat into the depth of the material (high thermal conductivity) *and* has the ability to absorb a lot of heat with little change in temperature (high thermal capacity), then the heat will be absorbed into storage within the material and there will be little rise in temperature at the surface. Hence, reradiation to other surfaces in the room and convection into the adjacent air will both be minimal.

Conversely, if the radiation is absorbed by an insulating surface with little thermal capacity, the surface temperature will rise rapidly, causing considerable reradiation to other surfaces and convection exchange with the room air.

It is clear that the selection of surface colour and the thermal properties are important when considering the dynamic behaviour of buildings. For convenience, the two thermal properties, conductivity and capacity, are multiplied to produce the thermal penetration coefficient, typical values of which are given in Table 5.1. Selection of materials with high thermal penetration coefficients (e.g. dense concrete, granite, limestone and slate) causes maximum absorption of heat, whether by conduction, convection or radiation, and materials of low thermal penetration coefficients (e.g. foamed lightweight concrete, cork, gypsum plaster, slag-wool and wood) cause minimal absorption of heat.

The source of heat, whether it be the sun, heating services, people or processes, only affects the manner in which the heat initially behaves in the space. For example, a timber park bench is warmer to the touch after a few minutes of contact through sitting than a stone wall, even though they may have been at the same temperature initially. Similarly, radiation penetrating into a building with timber floors and lightweight plastered walls, creates warmer conditions after a few minutes than the same amount of radiation penetrating an all concrete building.

However, consideration must also be given to the thickness of the material (one extreme being glazing for which the thermal capacity is ignored) and also whether the surface may be covered. For example, the absorbing properties of a dense concrete floor will be diminished if the floor is covered by an insulating carpet. As the thickness of the insulating covering is increased so the influence of the denser backing material is delayed and diminished.

The prediction of the thermal performance under dynamic conditions can be obtained using the admittance procedure (see Section 5.2.6) but nowadays it is more usual to use digital computers for dynamic simulation studies (see Section 6.3) which can deal with multilayered materials easily (e.g. carpet on underlay on cement and sand screed on concrete on hardcore on subsoil).

5.2.4 The Influence of Internal Activities

There may be incidental heat sources influencing temperatures inside buildings, such as people, animals, appliances and production equipment, which may generate heat continuously or sporadically (see Section 3.5). Their influence can range from insignificant (e.g. in warehouses) to highly significant, such that the internal temperatures are dominated by the internal activities (e.g. in foundries).

On some occasions the internal activities cause cooling, either by evaporation (e.g. water spray washing processes) or by cold process materials passing through the building (e.g. water works).

The effects of convection and radiation on the internal temperatures can be surmised from previous discussions. However, it should be appreciated that as surface temperatures and areas increase, so the strength and temperature of convection currents increase. Some industrial processes with large areas of very hot surfaces (e.g. foundries) cause convection currents which are hot, strong and turbulent, and these can dominate the internal air movement patterns and air temperature gradients. Strong radiation from large hot surfaces can also cause discomfort which may require control by insulation and shielding.

Each type of heat source must be considered individually for its effects on convection and radiation in the space and the influence on the internal temperatures.

The internal activities may cause movement of the air through physical movement of, for instance, production lines, conveyor belts, people and equipment. The resulting air movement will cause some stirring of the air in the space and consideration must be given to whether this is beneficial (e.g. reducing horizontal or vertical temperature gradients) or detrimental (e.g. spreading atmospheric pollutants into previously unpolluted areas).

5.2.5 The Influence of Ventilation

The effect of ventilation depends on the difference in temperature between the outside air and the inside air, and on the amount of ventilation as a proportion of the room volume, i.e. 'room volumes per hour' (Section 2.4.3).

If the outside air is cooler than the inside air, it will cause heat energy to be removed from the building when the cool air is introduced and the warm air is exhausted; the opposite effect occurs when conditions are reversed. A common application is to ventilate excess solar heat collected through glazing (see Section 3.5.5). However, because the intensity of solar radiation is constantly changing, so the ventilation rate must change constantly to maintain an even internal air temperature.

Ventilation by opening windows (Section 4.4 and 4.6) can work well in removing heat when the external air temperature is sufficiently below the required internal air temperature but not so low that discomfort is caused by temperature variations. However, excess solar heat cannot be exhausted when the external air temperature is the same as the internal air temperature and alternative methods of removing the heat must then be used (e.g. mechanical cooling).

5.2.6 The Prediction of Temperatures in Buildings

If all heat flows remained constant with time, there would be little problem in predicting the internal temperatures, but as external air temperatures and solar radiation intensities are constantly varying, the method of prediction must take into account these variations in order to achieve a reasonably accurate result.

Modern techniques for predicting temperatures are introduced in Chapter 6, from which it will be seen that one of the basic techniques for energy predictions using digital computers originates from the calculation of the temperatures inside buildings as time progresses.

One of the traditional methods for calculating transient temperatures by hand is the admittance procedure [5.11, 5.12] which was developed to enable the maximum temperatures to be predicted from diurnal changes. It requires data on external air temperatures and solar radiation intensities and this is used to calculate the transmission through each part of the opaque fabric and each orientation of glazing. The response of the room to these heat inputs plus any other internal sources of heat (people, lights, equipment, etc.) is then predicted from the ability of all the room's surfaces to absorb this heat, using 'admittance' or *Y*-values [5.12].

The calculation predicts the mean environmental temperature over the day in question and the environmental temperature at any selected time. Forethought and experience are required to determine at which time the peak environmental temperature will occur. As an alternative the same calculations can be performed by digital computers in less time.

Another method of predicting temperatures is to use the electrical analogy to heat flow, e.g. [5.13]. Heat transfer is simulated in electrical circuits by the flow of electrical current. Such systems are referred to as 'electrical analogues' or 'analogue computers' and for difficult problems these have sometimes proved to be the easiest method of predicting transient temperatures, especially when it is necessary to consider heat flow in three dimensions at once. However, most electrical analogue functions have now been replaced by digital computers (see Section 6.3).

Most temperature prediction methods for buildings simplify the actual heat transfer to a one-dimension problem which assumes that heat flows at 90° to the flat plane of any surface. Hence, the whole of a flat surface can only be simulated by one temperature and there can be no air temperature gradient in a room, only one air temperature value. Consideration must be given as to whether this simplification, and other simplifications and assumptions employed to render the calculations commercially possible, are sufficiently close to reality not to cause major discrepancies between the predictions and the temperatures which would actually be measured. These aspects are discussed further in Section 6.3.

Research and experience have demonstrated the value of the admittance procedure as a simple manual process which produces answers of sufficient accuracy for certain commercial, educational and domestic applications. It is frequently used when quick estimates are required during building design, e.g. [5.14].

5.3 ACTIVE METHODS OF TEMPERATURE REGULATION

5.3.1 Introduction

The previous section has discussed the scope for temperature regulation by the passive use of the building, windows, shading, internal activities and ventilation, but often it is not possible to achieve the desired temperature passively or to maintain it for the required period of time. It will then be necessary to consume energy to control the internal temperature by active methods.

If passive control fails to reach the desired temperature, then heat energy must be supplied to raise the temperature (i.e. a heating system). Conversely, if passive control exceeds the desired temperature then energy must be used to remove the excess heat energy from the building (i.e. a cooling system). If passive control fails both ways, then both heating and cooling systems will be required.

Heating can be achieved by increasing the air temperature, increasing the radiant temperature or a combination of the two. Cooling is usually achieved by reducing the air temperature.

The speed and nature of response of the internal space to imposed temperature changes caused by mechanical services depends on the factors already described when discussing temperature changes due to solar radiation penetrating glazing (Section 5.2.3). However, the heating system does affect the nature of heat transfer in the space (e.g. by natural or forced convection, or by radiation) with further implications on aspects such as comfort and control. The heating system itself also has inherent response characteristics (see Section 6.5.1) which affects its performance during intermittent operation and when responding to varying loads.

The selection of the most appropriate form of temperature regulation system and its correct location in the controlled space are both vitally important in order to ensure efficient control of temperature and therefore efficient utilisation of the energy consumed by the system. Incorrect selection and location are two of the major causes of poor temperature control and excessive energy consumption.

It will be apparent from the following discussion that the selection of a position for a piece of equipment or the position where conditioned air is delivered to a space depends on how the system is intended to control temperature and the location where heat is arriving or departing from the space.

5.3.2 Heating by Increasing the Air Temperature

Air will naturally be more buoyant when heated and so there is the inherent problem of maintaining heated air in the required zone, such as the occupied zone in high bay warehouses. This effect is reduced by heating a larger quantity of air to not such a high temperature and this reduces the temperature of the stagnant layer of warmed air adjacent to the ceiling or roof. Consequently, the heat lost from these upper areas is also reduced.

The variation of air temperature with height is often referred to as the vertical air temperature gradient. In addition to energy conservation, the control of the vertical air temperature gradient is also necessary to prevent thermal discomfort in tall spaces which are occupied throughout the full height (e.g. mezzanine floors in a machinery hall).

Vertical air temperature gradients will form part of the following discussion on air heating systems. The design of the heating system usually aims to minimise the temperature gradation between high level and low level by the correct selection of heating equipment, instead of installing remedial air distribution equipment to rectify excessive vertical air temperature gradients. However, occasionally vertical air temperature gradients are desirable, such as in saunas where the variation in temperature with height caters for occupants of differing experience and tastes, and in cooling applications where the stock or the occupants require low air temperatures just at a low level (Section 5.3.8).

Air heating systems can be classified into the following three groups.

(*a*) *Natural convector heaters.* Air flow through convector heaters occurs by the stack effect only (Section 4.6). The heated air is allowed to leave the heater from the top and to enter the space, after which it will continue rising unless influenced by stronger air movement, or it meets an obstruction. Consequently, convector heaters must be positioned as low as possible in the space in order to heat the whole height of the space by convection air circulation.

The stack effect within convector heaters is relatively weak and so generous areas must be provided by grilles in the heater cases for air ingress and egress (Fig. 5.3). Even so, the majority of convector heaters discharge the air at relatively high temperatures (e.g. 105 °C for some electric convectors) in order to obtain greater buoyancy and so reduce the size and cost of the casing for the heater. This can lead to considerable overheating

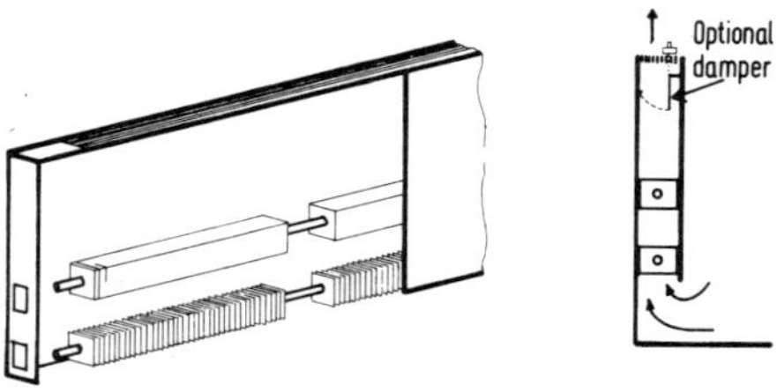

Fig. 5.3. A typical natural convector heater with an optional manually controlled damper for reducing the heat emission by restricting air movement. The casing is shown partially removed to expose the finned tube heaters.

of the space above occupancy level by the hot rising convection currents. Typically, in a heated office, air temperatures adjacent to the ceiling can be 32 °C when the air temperature adjacent to the floor is only 18 °C. At head height the air temperature might be 21 °C. Hence heat transfer through the ceiling and roof is increased by the unnecessarily high localised temperature, leading to considerable energy wastage.

Filters cannot be incorporated into natural convector heaters as they cause too much resistance to air flow but, in non-electric heaters, manually controlled dampers are often provided to reduce heat output by preventing air movement through the heater. Automatic control of heat output can be achieved by varying the temperature of the air heating fins or by on/off control of the heating element. Heat can be supplied by steam, water or electricity and as there are no moving parts the heaters do not produce noise. Heat can also be supplied by the combustion of gaseous fuels within a combustion chamber, provided that fresh air can be supplied for combustion and the combustion products can be safely ducted away through a flue.

The heaters are available as completely manufactured items or in component form for installing in purpose-designed casings within the building.

(*b*) *Fan assisted convector heaters* (*'fan heaters'*, *'unit heaters'*). By using a fan, the air flow rate through a convector heater can be increased. This can result in benefits such as reduced discharge air temperature, far better control over the direction of the discharge air and reduced size of the heater. Frequently, filtration is incorporated to prevent impaired performance when dust and dirt collect on the fins of the heat exchanger (Fig. 5.4).

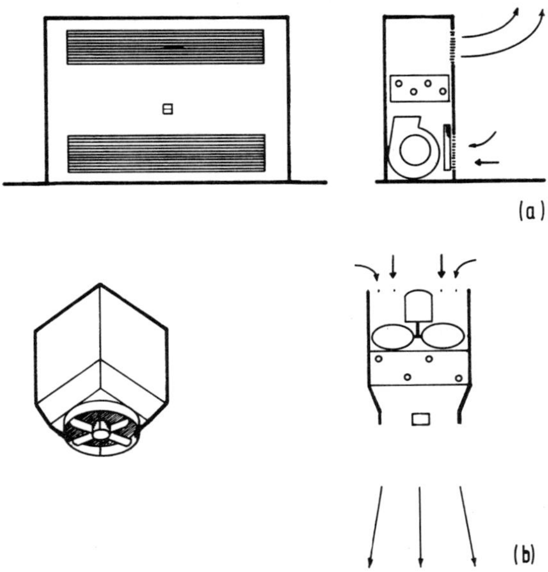

Fig. 5.4. (a) A typical fan assisted convector heater showing air movement through the filter, fan, heater and discharge grille. The switch located in the centre of the front casing can be used to turn off the fan. (b) A typical unit heater illustrating air movement through the fan, heater and radial discharge vanes. The view from below (on the left) shows the radial discharge vanes which must not be adjusted to deflect the warm buoyant air jet too much, otherwise it will not penetrate the colder denser air below to hit the floor.

Normally, the fan assists the natural air flow direction but it can be used to force air through the heater in another direction. Hence, downward discharge of the heated air is possible from heaters located above the occupied space.

Fan assisted convector heaters are usually controlled from a temperature sensor to achieve on/off operation of the fan or to vary the fan speed, but the heater temperature can be modulated if cool or cold discharge air is acceptable in the space.

Because of the moving components, the heater produces noise and will sustain wear. Maintenance will be necessary to clean filters (if incorporated) and occasionally to attend to the fan motor bearings or the electrical supply.

Fan assisted convector heaters can be purchased as manufactured items for operating from steam, water or electricity. They can also be obtained in carcase form for installation in purpose designed casings within the building.

(*c*) *Warm air systems.* Warm air systems remove the air from the space and duct the air to a convenient location where a fan can be housed and equipment can treat the air. The air is then returned to the space via ductwork (Fig. 5.5).

In addition to heating the air, the system can provide filtration to high standards, humidification and other specialist treatments. The noise generated by the system can be attenuated by acoustical absorption material ('silencers'). If cooling is incorporated, the system usually becomes

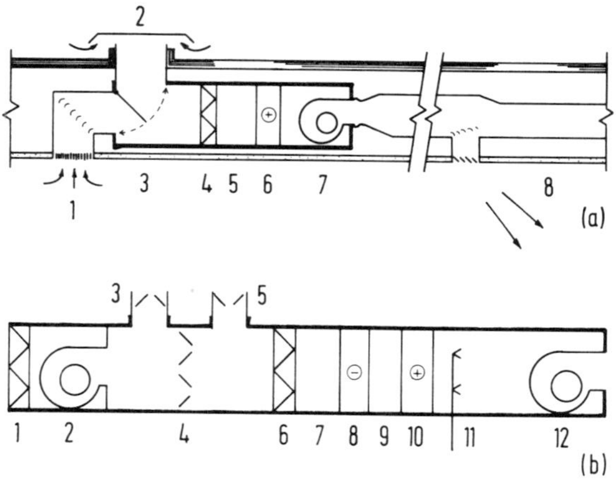

Fig. 5.5. (a) A section through a simple packaged warm air heating handling plant located within a ceiling void and incorporating: 1. Return air grille and ductwork, 2. fresh air inlet, 3. damper to select the proportion of fresh air to recirculated air, 4. filter, 5. access section, 6. heater, 7. fan, and 8. distribution ductwork and supply air diffusers. Access through the ceiling void must be provided for maintenance, particularly to replace dirty filters. (b) A section through a typical air handling plant with additional facilities to include: 1. Return air filter, 2. return air fan, 3. exhaust air dampers and duct connection, 4. recirculation air dampers, 5. fresh air duct connection with dampers, 6. supply air filter, 7. access section, 8. cooling coil (see Section 5.3.5), 9. access section, 10. heating coil, 11. humidification facilities (e.g. steam injection), and 12. supply air fan.

larger and the process is often called 'air conditioning' (see, for instance, [5.5, 5.15, 5.16]).

Dehumidification can also be incorporated and when complete control of temperature and humidity are both provided the system is often called 'full air conditioning'. However, these terms are sometimes used glibly and so may need to be treated with caution.

Each warm air system is designed for each application. Hence, return and supply air positions can be selected as appropriate to the general air flow requirements, the supply air temperature can be designed for acceptable buoyancy and vertical air temperature gradients under full load, and the system can be used to influence spatial air movement patterns by the location of the supply and extract positions and by the characteristics of the supply air jets (see Section 4.2).

The principal system components can be obtained in modular form which facilitates interfacing fans, filters, heating coils, etc., and such systems are called air handling units. The duty of each component is selected for the particular application.

Heat can be obtained from steam, water or electricity, or any combination. Alternatively, the air can be heated indirectly (through a heat exchanger) using gas or oil fired burners, provided the combustion products can be expelled easily through flues. Often, indirect fired units are mounted externally and the air is ducted to and from the occupied spaces. Maintenance will be necessary to clean filters, to replace fan belts, to adjust fan belt tension, and to clean and adjust the burners. Control is achieved by variation of the heater temperature and the fan usually operates continuously during the period when space temperature control is necessary.

When the combustion of a fuel produces no toxic products, the fuel is sometimes fired directly into the ducted air stream. Whilst this form of heating usually achieves the highest efficiency, consideration must be given to the combustion products (carbon dioxide, water vapour, etc.) to determine whether the resultant pollution of the air in the space is acceptable. This form of heating requires careful safety controls and continuous ventilation to prevent the concentration of combustion products becoming excessive. Hence, their application is usually restricted to buildings which require high ventilation rates, e.g. factories.

The amount of air which passes through warm air systems is often quoted in terms of the air recirculation rate, which may be expressed in terms of room volumes of air per hour (RVH). Consideration must be given to the resulting temperature of the air supplied to a room for a given recirculation rate. As the supply air temperature increases, so the recirculation rate can be reduced to meet the same duty, but the supply air is then more buoyant and this worsens the problem of the air temperature gradient, unless the air in the space is well stirred. There is always a cost incentive to reduce the recirculation rate because most components in the system will then be smaller and so the capital cost will decrease but this is limited by the need to prevent excessive supply air temperatures.

The maximum permissible supply air temperature has to be determined for each application and system design, having regard to the spatial requirements. On some occasions when careful control of air temperature is required in a space, the maximum supply air temperature might be restricted to only 5 °C above the space temperature. This limits horizontal and vertical air temperature variations under maximum heating and also causes little change to the room air circulation patterns between the extremes of buoyant discharge at full heating to no buoyancy of the supply air when there is no heating.

The system can also be designed to introduce fresh air; this will often require a separate fan and filter. The amount of fresh air supplied can again be expressed in terms of room volumes per hour (RVH) (Section 2.4.3). This sometimes leads to confusion between the recirculation rate (which is essentially governed by heating, cooling or air movement requirements) and the ventilation rate (which is strictly the 'fresh air' rate for 'ventilation' purposes).

The warm air system return air ducts can also be used for the exhaust ventilation air and usually a separate return or exhaust air fan is necessary.

5.3.3 Radiant Heating

The radiation exchange equations (see Section 1.3.4) have shown that the rate of radiant heat transfer increases considerably with increasing absolute temperature and so it is high temperature heat sources which produce high radiation per unit area. Radiation will spread from the source but, with high temperature and medium temperature sources, the sources are relatively small and reflectors are often employed to focus the radiation in particular directions. These reflectors sometimes duplicate as shields to reduce convective heat losses from the heat sources. The fall-off of radiation intensity with distance from point, linear and planar sources is discussed in Section 1.3.7.

Radiation is only beneficial when it reaches the object to be heated. As radiation travels in straight lines the objective is to obtain a direct path between the source and the receiver, otherwise, as with light, radiation is weakened at each reflection from surfaces depending on the extent to which the surface absorbs the radiation (i.e. the absorptivity). For instance, a person sitting at a table in a room heated by overhead radiant heaters will have the lower part of the body shielded by the table top. As the legs and feet will not receive direct radiation, discomfort may be experienced.

In applications for thermal comfort, the difference between radiation intensities across the body (head to foot, front to back) must be limited when the occupants face one direction for long periods, otherwise discomfort will result (Section 2.3.7).

Nevertheless, radiant heating can be designed to produce comfortable conditions which can also lead to significant energy savings over air heating systems in generously ventilated buildings. This benefit derives from the lower air temperature at which comfort is experienced when radiant heating is employed (Section 2.3). Because of the lower air temperature, less energy is ventilated to waste in the form of warmed air (Section 3.3.4).

Radiant heating systems aim to maximise the radiation heat transfer, sometimes in preferred directions, and to minimise the convective heat transfer as generally this is of no benefit. Table 5.3 gives an indication of the variety of systems available, some of which are illustrated in Fig. 5.6.

Except for heated ceilings at around 30 °C, the temperature of the sources necessitates their positioning out of normal reach. Consequently, the radiation arrives at the receiver from directly above or from the sides above

Table 5.3 A selection of radiant heating systems

Temperature grouping	Type	Heat source
Low temperature sources (up to 120 °C)	Ceilings	Electricity Hot water
	Individual panels	Electricity Hot water Steam
Medium temperature sources (non-luminous—less than 500 °C)	Individual panels	Electricity Steam
	Tubes	Electricity Gas (indirect fired with flues) Steam
High temperature sources (luminous—up to 1 000 °C)	Panels, strips and tubes	Electricity Gas (direct combustion products enter the space)

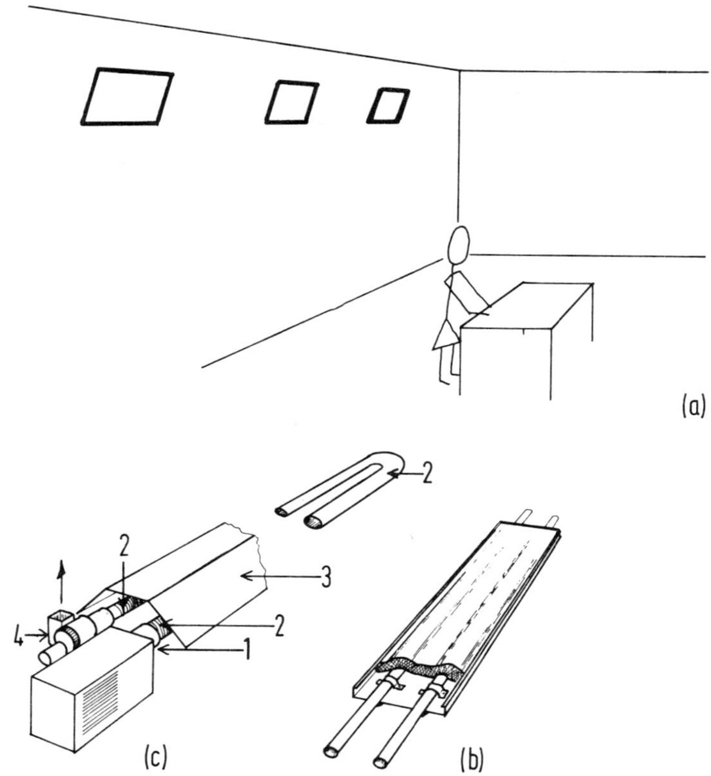

Fig. 5.6. Examples of radiant heat emitters: (a) Medium temperature wall mounted radiant panels; (b) medium temperature over-head panel viewed from above, illustrating the insulation cut back to expose the steel panels and tubes; (c) high temperature radiant tubes with polished metal reflectors viewed from above. Gas and air are mixed and ignited at the burner (1) and the resulting combustion heats the tube (2), emitting radiation in all directions. The polished metal reflector (3) directs heat downwards. A fan (4) draws the combustion products from the tube (2) and discharges them to waste through a flue (not shown).

occupancy reach and so sufficient height must be achieved above occupied spaces to prevent uncomfortable differences in radiant intensities to the head and the feet. Recommended radiation intensities are given elsewhere (e.g. [5.5, 5.17]).

5.3.4 Combined Radiant and Air Heating

Systems which operate by both radiation and convection involve a combination of the characteristics which have already been discussed.

Briefly, for efficient use of the convection currents, the systems must be at low level in the space, but also located where the radiation can be of benefit. Because they can usually be touched, the surface temperature must be limited to prevent burning, which in turn limits the amount of radiation. In fact, flat single panel 'radiators' against a wall provide approximately 40 % of their heat output by radiation and this proportion falls as fins or extra panels are added to increase the convective heat emission (e.g. [5.18]).

There are a variety of systems designed to produce useful heating by radiation and convection. These include:

Radiators—hot water, oil filled electric, dry electric
Storage heaters—off-peak electricity.

5.3.5 Cooling Systems

The effectiveness of radiation heat transfer decreases significantly as the absolute temperature falls (Section 1.3.1). This phenomenon restricts the design of cooling systems to convective techniques. Forced convection is usually employed in order that maximum heat transfer is obtained, as the difference between the supply air temperature and the mechanical cooling system's temperature is smaller than the equivalent temperature difference in a heating application.

The control of the air temperature in a space is usually based on varying the supply air temperature from the air handling system (Fig. 5.5(b)). When it is necessary to cool the space, the supply air temperature is reduced by cooling the air as it passes through the cooling coil, which is itself fed with chilled water or refrigerant from a mechanical cooling system.

A more complex design for cooling systems is based on varying the air quantity as well as the supply air temperature, because the energy consumed by the fans can be significant [5.24, 5.25]. By halving the air volume, for example, the energy consumed by the fans falls to one quarter.

The majority of mechanical cooling systems use vapour compression refrigeration machines, as illustrated in Fig. 5.7. Briefly, a vapour (called the refrigerant) is compressed which raises its temperature. At a convenient location (e.g. the external air) heat is rejected to cause the refrigerant vapour to cool and condense into a liquid, after which the liquid refrigerant is piped to wherever cooling is required. Here, the refrigerant is forced through a restricting valve and, at the much lower pressure, the refrigerant boils and takes its latent heat of vaporisation from its surroundings. The refrigerant vapour is then drawn back to the compressor where it is recompressed to form hot refrigerant vapour and the cycle is repeated. So the refrigeration

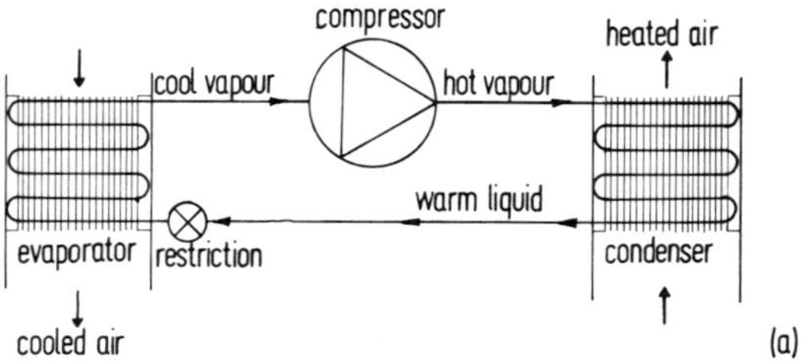

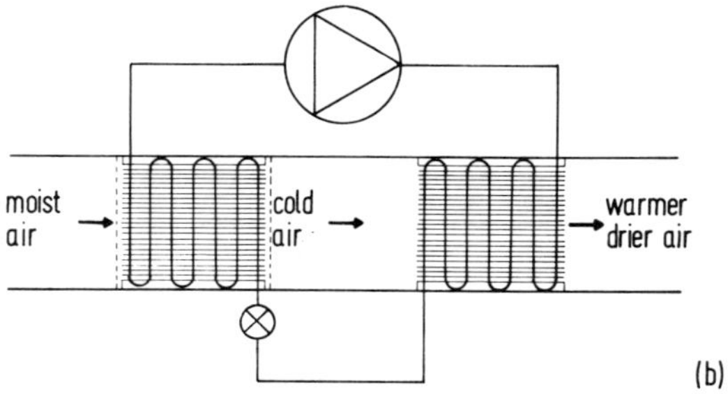

Fig. 5.7. Principal components of vapour compression refrigeration machines. (a) Application in which heat is transferred from one air stream to another. (b) Application in which heat is returned to the same air stream after dehumidification. A condensate collection tray and drainage system must be located under the evaporator (shown dotted on the plan).

machine has taken heat from a source (where the refrigerant evaporates) and this heat, plus the energy required to force the refrigerant around the system, has been rejected at a higher temperature. When this heat is utilised the system is called a *heat pump*.

The amount of energy taken in at the evaporator, i.e. the refrigeration effect, is usually greater than the amount of energy used by the compressor to obtain the refrigeration effect. The ratio of energy removed at the evaporator to the energy input at the compressor is called the coefficient of performance (COP). As an indication only, the best performance of around 6 to 1, when the refrigerant evaporating and condensing temperatures are similar, falls to around 2 to 1 when there is a wide difference in temperature (e.g. 50 °C). However, there are many factors which influence the performance of refrigeration equipment (e.g. [5.19, 5.20]) and so refrigeration specialists should be consulted.

The rotary power requirement at the compressor can be supplied by an electric motor or a diesel or spark ignition engine fuelled by liquid or gaseous fuels (e.g. oil, petroleum spirit, liquified petroleum gas (LPG), methane or natural gas). The overall efficiency of the primary fuel utilisation, called the coefficient of fuel utilisation (CFE), should be considered in addition to the refrigeration machine's COP. This will enable comparisons to be made between the amount of refrigeration obtained (or heat rejected) compared to the total primary fuel consumption, including obtaining the fuel from its source, processing, distribution, (and in the case of electricity, conversion to electricity and transmission losses) and, finally generation of the motive power for the compressor (e.g. [5.21]). Engine driven compressors provide the opportunity for recovery of the high grade waste heat from the engine, especially useful when the system is a heat pump. The equivalent heat in the electrical system is rejected at the electricity generating plant.

Most projects of a small nature use the refrigerant directly and this application is referred to as direct expansion. On large projects or where there are many different air temperature requirements, the refrigerant is interfaced with water systems to either collect the heat for the refrigerant or to reject the heat from the refrigerant or both. These are called chilled water and condenser water systems respectively. Brine or other antifreeze is used in chilled water systems when the temperature must be taken below the freezing point of fresh water.

When applied to the cooling of air in buildings, the temperature of the refrigerant in air cooling coils is restricted by the problem caused by freezing the water vapour condensed from the air. Special provision must be made for finned cooling coils which are intended to operate at temperatures below freezing point as ice will form on the coils and air flow through the coil will diminish. Defrosting or de-icing facilities must be incorporated otherwise a solid block of ice will form and no air flow will occur. However, such low refrigerant temperatures are only used when absolutely necessary because the refrigeration machine's coefficient of performance decreases as the difference between the condensing temperature and the evaporating temperature increases. Hence, in order to minimise energy consumption, the evaporating temperature is kept as high as possible, as the condensing temperature is usually fixed by the temperatures where the heat is rejected (e.g. external air temperature in summer). But there are other engineering considerations which cause the selection of temperatures to be a compromise and so cost optimisation techniques are necessary.

One opportunity for saving energy exists in air conditioning applications when it is not necessary to dehumidify the room air. The refrigeration machine's evaporating temperature can be kept above the room's dew point temperature so that firstly, the coefficient of performance is raised, and secondly, energy is not consumed by the refrigeration machine for condensing moisture from the room air. Consequently, air removed from an occupied space at say 21 °C and 50 % RH can only be cooled to a minimum of around 12 °C, depending on comfort requirements, at which temperature it is close to its dew point. This temperature difference of 9 °C can be compared to that which might occur for ventilation only plant achieving the same duty. If the ventilation plant is designed to provide cooling for a temperature difference of half that of the cooling system, i.e. 4·5 °C, it will handle twice the air quantity, ducts and fans will be approximately twice the size, and even so it will not achieve the required duty when the temperature difference is less than 4·5 °C, i.e. the external air temperature exceeds 16·5 °C (21 °C–4·5 °C = 16·5 °C).

Consequently, for the majority of commercial applications the difference between the room air temperature and the cooling coil can be little more than 9 °C. This can be compared to the 50 °C difference with conventional radiators in heating systems, emphasising why cooling requires the better heat transfer achieved with forced convective rather than natural convective techniques.

With most cooling systems, condensation will occur occasionally on the cooling coil, even when dehumidification was not intended, and so condensate drains are often installed. Some individual room cooling units simply collect the condensate in trays from which it re-evaporates into the room only to be condensed on the cooling coil again later.

In situations where dehumidification is deliberately intended, the air is cooled well below its dew point to condense much of the water vapour from the air and the condensate is piped away. Then the air is reheated to the required temperature. Sometimes the hot refrigerant vapour can be used to reheat the air and this avoids consuming more energy for the heating function (see Fig. 5.7(b)).

The minimum supply air temperature for a space must be considered in a similar way to that in which the maximum supply air temperature was considered for the supply of warm air from heating systems. In occupied spaces the supply of cool air must not cause discomfort through horizontal temperature variations, although a vertical air temperature gradient in tall spaces can be accepted when the temperature above the occupied zone is of little concern. The location of both the cool air discharge and the extract can then be positioned so that only the occupied zone is cooled, as illustrated in Fig. 5.8.

Cool air which reaches the occupants before entrainment of the room air has sufficiently tempered the chilling effects of the supply air jets, will give rise to complaints of discomfort. Hence, the selection of the supply air

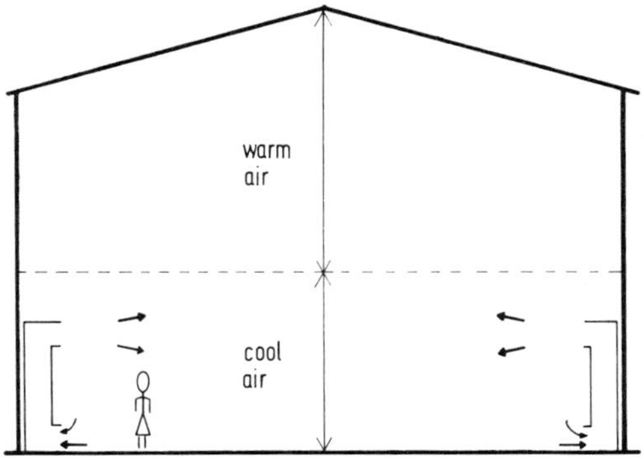

Fig. 5.8. Possible arrangements for efficient cooling of tall spaces where low temperature is required at low level only. The air discharge arrangements should minimise the disturbance of the stratified warm air above the occupied zone, yet discharge just above the occupied zone is desirable to avoid discomfort caused by cold draughts.

temperature, the spacing and direction of the supply air discharge jets and the velocity of the discharge, are all related to the height of discharge and comfort considerations. Having determined the supply air temperature, it is then possible to calculate the quantity of air which must be handled by the system, although as before there is a capital cost saving by minimising the supply air temperature as this reduces the size of the majority of the components in the system. However, the energy consumed by the refrigeration plant increases and so the running costs increase. Hence, there is a cost optimisation exercise to balance the capital costs and operating costs.

The refrigerating equipment can either be mounted within the space (unit air conditioners) or the air can be ducted to a location where it is more convenient to situate the equipment (air conditioning installation). The refrigeration compressor can either be mounted within the equipment or located remotely. Large installations usually have the compressors located remotely (Fig. 5.5(b)). The control of noise is another important consideration when siting the refrigeration compressors.

The control of a large cooling installation can be quite complex and details can be found elsewhere (e.g. [5.15, 5.22]). For a small installation, the compressors can simply be turned on or off by a control system or sometimes controlled in stages, depending on the facilities available and the accuracy of temperature control required. When chilled water is used, the control can be fully modulating using motorised valves and this provides the facility for more accurate temperature control.

5.3.6 The Control of Humidity

The relative humidity of fresh ventilated air entering a building may need to be controlled. In cold conditions, heated fresh air will have a low relative humidity, whereas in hot summer conditions outside air may be at a high temperature and a high relative humidity. The usually quoted relative humidity range within buildings for human comfort and control of mould growth is 40–60 %. Some industrial or commercial applications require more strict control of relative humidity, e.g. 50 % ± 5 %. Examples are printing processes, where the dimensional stability of the paper is dependent on the relative humidity, in computer suites, where the performance of electronic equipment is affected by relative humidity, and in the storage of corrodable metals when the relative humidity should be kept below 50 %.

Humidification can be achieved by a variety of methods. For example, an atomising water sprayer takes heat from the air for the latent heat of vaporisation to evaporate the atomised water into water vapour. Consequently, the humidified air is colder and denser than the room air and will tend to fall after discharge into the space. Alternatively a steam humidifier causes a negligible change to the air temperature after the steam has been injected into the air but the density of the air decreases slightly due to the presence of the water vapour and so it will tend to rise after release in the space. This effect is exemplified by kettles left boiling in a kitchen—the water vapour remains predominantly at high level unless the air in the space is stirred.

In addition to atomising humidifiers and steam humidifiers, humidification can also be achieved by straightforward evaporation from pans or pans and wicks. Both of these techniques take the latent heat of evaporation from the air, which naturally becomes cooler and denser. If heating is also required and is incorporated before the humidifier, then the cold dense discharge can be avoided [5.15, 5.16, 5.22]. Cleanliness is essential to control allergens (humidifier fever) [5.26] and disease (e.g. Legionnaires') [5.27].

When air is dehumidified there may also be a change in the density. For example, heated desiccant dehumidifiers can cause an increase in the air temperature, resulting in reduced density and therefore increased buoyancy of the discharge air. Dehumidification by refrigeration processes initially causes an increase in air density due to the cooling of the air but this is often followed by a heater or the refrigeration machine's condenser which rejects the heat previously taken out of the air (Section 5.3.5). This results in increased buoyancy, depending on the amount of heating.

When an air handling system is controlling both temperature and humidity, it can be seen that the trajectory of the supply air jet will vary according to how the air temperature and humidity are each varied. In spaces in which the air is well stirred the buoyancy variations will be of no concern, but when dealing with large spaces, or spaces with minimal air movement, consideration of the buoyancy of the discharge air is most important to ensure that the conditioned air is delivered to the required location. Only then will the air temperature and humidity be controlled well.

On many occasions, consideration of the control of air temperature and humidity does not receive priority in the overall design of the space. The design may be compromised by overriding interests such as aesthetics, cost, availability of space and access for maintenance. Hence, if it is accepted that a new building is to be designed to achieve a high standard of energy control, then the location and operational characteristics of the services must receive high priority.

5.3.7 The Positioning of Heating or Cooling Equipment

In the previous sections dealing with the performance of temperature regulation equipment, it has been stated that the location of equipment, or the location for the supply of heated or cooled air, is important to ensure efficient control of space temperatures and therefore energy consumption. Further consideration of the activities, solar heat gain and heat loss is now necessary to achieve the best arrangement.

When considering the heating of buildings, heat is lost through the envelope and through the floors when they are in contact with the ground, external air or unheated space. In order to maintain a constant and even air temperature it is only necessary to provide heat in the areas of heat loss and fresh air ingress, and then only sufficient to equal the amount of heat lost.

Uncontrolled cold downdraughts from cold surfaces such as windows will fall to the floor where they will spread out, causing sedentary occupants close to and some distance from the cold surface to suffer discomfort through cold feet and ankles. This problem can be resolved by placing convective heating equipment beneath cold surfaces so that the buoyant convection currents oppose the denser cold convection currents falling from the cold surfaces (see Fig. 5.9). The objective is to arrange for the two opposing convection currents to be approximately equal in strength; rising convection currents which are too strong will cause additional heating of

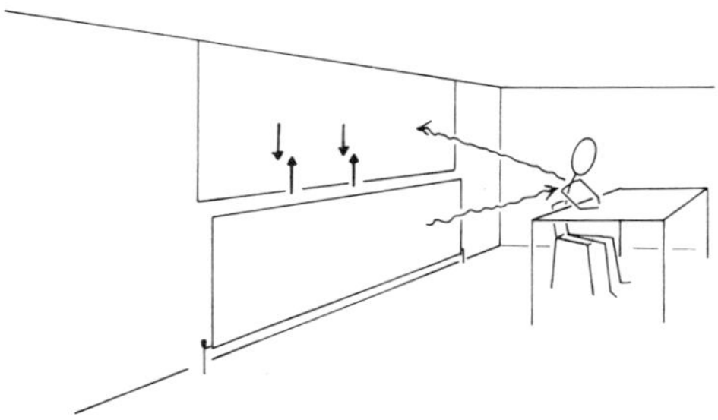

Fig. 5.9. Radiators located beneath windows aid comfort: warm convection currents rising from the radiator oppose the cold falling currents from the glazing, and radiation heat transfer to sedentary occupants balances their radiation heat transfer to the cold glazing.

the cold surface; exacerbating heat loss and energy consumption, while rising convection currents which are too weak may be dominated by the cold downdraughts, causing vertical air temperature gradients and thermal discomfort. Occasionally the cold currents and warm currents will move in columns adjacent to each other when they are each of widely differing temperature to the spatial air temperature. Nevertheless, it is common to see convective heat sources placed under single glazed windows especially when tall windows cause strong cold convection currents.

Although the heating of cold surfaces can eliminate the cold downdraught problem and improve the radiant temperature, these improvements to thermal comfort are obtained at a substantial increase in energy consumption. The reduction of cold downdraughts and the improvement of the surface temperatures can both be achieved passively. For example, the improvement can be achieved by insulation of the building envelope, or in the case of windows by double or triple glazing and by covering the windows internally with curtains or blinds and externally by blinds or louvres. The now weaker cold downdraughts can still be opposed by correctly sized convective heating placed beneath the cold surfaces, thereby achieving thermal comfort and good temperature control at a minimal consumption of energy.

These cold air currents can sometimes be warmed when they reach the floor by underfloor heating systems or by skirting convectors placed around the periphery of the floor. However, these techniques are generally not so effective for sedentary occupants who are situated close to the cold surfaces.

The placing of radiation heat sources can also help to even the temperature across the space. For example, sedentary office staff may find discomfort by radiation heat loss from one side of their body to cold surfaces such as single glazing. They may complain of cold draughts because the cause of discomfort may not be apparent to them but a study of air temperature and air movement will often indicate that cold draughts are not present in the vicinity of the complainant. This problem may be alleviated by placing radiation heat sources in the vicinity of the cold surface and directed towards the occupants to compensate for the radiation loss from the occupants to the cold surface. Hence, it is common to see radiators under windows as they provide a contribution to radiation balance for comfort, as well as helping to oppose the cold convection currents (see Fig. 5.9).

When faced with sources of heat in a space, such as hot industrial plant, there is the option of either treating the excess heat as a useful contribution to heating the remainder of the space by redistribution of the heated air, or treating the heat as an unwanted pollutant which must be removed from the space for use elsewhere or discarded. The requirement may vary with the time of year.

The process of removal normally requires an air handling system to collect the warm convection currents rising from the hot plant using the principles of capture discussed in Section 4.2.3. The extract should be located over the plant and collection efficiency may be improved by the use of hoods, canopies and screens. Having collected the heated air, the system can then either reject the heat from the space or make use of the heat elsewhere in the space or in other spaces. In this case the location of the supply of the warm air must conform to the principles discussed in Section 5.3.2 to ensure even air temperatures throughout the space.

When capturing hot convection currents from many individual hot surfaces, the principles of capture are applied to each surface. Unfortunately, these convection currents are not always conveniently grouped together and when, for example, solar radiation is penetrating multiple windows in an open plan office it is clear that the relatively much weaker convection currents will originate from wherever the tracking radiation is currently landing (desks, carpet, filing cabinets, etc.). In such situations the control of air temperature is more easily achieved by ensuring that the air in the space is well stirred so that there are no significant spatial air temperature variations.

Stirring can be created by room fans or more commonly by the direction and velocity of the supply air jets of any air handling system. These must create just sufficient room air movement to mix the rising convection currents with the room air without causing draughts or spreading other pollutants into clean zones.

When the air is sufficiently stirred, air temperatures will be similar through the space and so the location of an air handling system's return air extract point is immaterial, provided the system does not remove the supply air immediately after entry to the room or before its function has been completed (i.e. 'short circuiting').

For convenience, return air grilles are often mounted overhead although, when an air handling system supplies cool air, the return air grilles can be located at low level to deliberately restrict the cooled air to low level (see Fig. 5.8). This will reduce the running costs but consideration must still be given to the hotter conditions in the space above, as this may cause radiation gain to the occupants or products below.

When an air handling system supplies cool air on some occasions and warm air on other occasions, ideally it should have two supply arrangements to discharge air in the direction appropriate to its buoyancy and possibly two extract systems at different levels in the space. Needless to say, such designs are not common as the design of the services is usually arranged to compromise between the two extremes of buoyancy; the engineer having first checked that the performance at the two extremes is adequate for the requirements of the space. Hence, it is clear that the location of equipment, and the orientation and location of air handling terminals, has to be a compromise between the extremes of buoyancy from maximum heating to recirculation with no buoyancy variation, and for cooling applications, to maximum cooling; possibly with further buoyancy variations caused by changes in the moisture content of the air.

5.3.8 Effectiveness of Controls for Temperature Regulation Services

Section 5.3 has discussed active methods of temperature regulation using mechanical systems to alter either the air temperature or the radiant temperature, or sometimes a combination of both, together with influencing air movement. Systems may also be incorporated for humidity regulation. The need for this regulation of temperature or humidity is sometimes for the process or contents and sometimes for human comfort, but whatever the application the important parameters to control are as described in Section 2.2.3, namely:

Air temperature
Radiant temperature
Air velocity
Relative humidity.

However, the mechanical installation is often selected to regulate specifically only one of these parameters. Furthermore, 'space temperature' sensors are usually designed to be sensitive to only the air temperature. This can lead to the illogical situation of radiant heating installations modifying the radiant conditions but being controlled by sensors which are only sensitive to the air temperature.

Ideally, for human comfort, the sensor should be sensitive to all four of the above factors, although the effect of relative humidity is insignificant if it lies in the usual range of 40 % to 60 %. Attempts have been made to produce comprehensive sensing devices and the eupatheostat is one [5.23]. This was designed by Dufton in 1929 to respond to air temperature, radiant heat and air velocity in a similar manner to the human body.

The velocity of the air over humans partially affects comfort by changing the rate of evaporation of moisture from the skin. This does not usually occur with materials because they are dry, and so an instrument for better environmental control for materials protection would have to respond differently to a eupatheostat. The air velocity is still important though to ensure adequate mixing of the air.

For efficient control, the sensor must be located where the controlled conditions have to be achieved. If the sensor has to be located elsewhere, then this is acceptable provided there is no change of temperature or humidity when the air leaves the controlled zone and arrives at the sensing zone, or if there is a change, then this change is caused by a constant influence (e.g. a known constant rate of heat input from production plant).

When there is more than one item of services for one zone, for example many fan assisted convectors providing heating and cooling in a long open plan office, it is common to have each unit controlled by its own thermostat situated behind the entering air grille. This is acceptable for comfort control but care must be exercised during commissioning of the controls to ensure that each control is set exactly to the same temperature, otherwise considerable amounts of energy will be wasted by some units on cooling whilst others are on heating. Furthermore, the commissioning should not rely upon the temperature indicator on the controls as this may not be calibrated sufficiently accurately. Provisions must also be made to prevent casual adjustments by the building occupants otherwise the benefits of a well commissioned installation will be lost. Even so, with time, some controls tend to go out of calibration due to natural phenomena and so recommissioning is necessary.

The problem of many controls in one zone can be resolved by having just one temperature sensor which controls all units. However, this leads to another problem; namely where the sensor should be located, particularly if there are local sources of heat (e.g. photocopiers) or different external influences (e.g. differing solar gains through windows). One solution is to ensure that the air in the space is well mixed so that heat released from equipment or solar penetration does not linger to cause local temperature variations. If one air handling system is serving such a zone, the temperature sensor is often placed in the return air duct where it will sense the air temperature after the air from every part of the zone has been mixed. Hence, the sensor is sensing an air temperature which may not be representative of the variety of air temperatures in the space, especially if there are considerable variations to the air temperature immediately before the air is extracted from the space. In such situations it is important to have sensors actually in the space and sufficient stirring of the room air to prevent localised temperature variations.

No attempt has been made here to cover details of the design of controls which may be found elsewhere (e.g. [5.15, 5.22]). However, some consideration has been given to the performance and location of control sensors as such factors are of great importance in the correct and efficient operation and running of any temperature regulation system. Design effort can be wasted, energy consumption can be unnecessarily high and many complaints from the occupants may be received if correct control is not achieved.

REFERENCES

5.1 BILLINGTON, N. S.: *Thermal Properties of Buildings*. Cleaver–Hume, London, 1952.

5.2 BILLINGTON, N. S.: *Building Physics: Heat*. Pergamon Press, London, 1967.

5.3 ADDLESON, L.: *Materials for Building. Volume 4: Heat and Fire and their Effects*. Newnes–Butterworths, London, 1976.

5.4 CHARTERED INSTITUTION OF BUILDING SERVICES: *CIBS Guide A6: Solar Data*. CIBS Publications, London, 1971.

5.5 AMERICAN SOCIETY OF HEATING, REFRIGERATION AND AIR-CONDITIONING ENGINEERS: *1977 Fundamentals Handbook*. ASHRAE, New York, 1977.

5.6 PETHERBRIDGE, P.: *Sunpath Diagrams and Overlays for Solar Heat Gain Calculations*. Building Research Station Current Paper, CP 39/65, 1965.

5.7 HARKNESS, E. L. and MEHTA, M. L.: *Solar Radiation Control in Buildings*. Applied Science, London, 1978.

5.8 GIVONI, B.: *Man, Climate and Architecture*. Applied Science, London, 2nd edn, 1976.

5.9 BURBERRY, P.: *Building for Energy Conservation*. Architectural Press, London, 1978.

5.10 VAN STRAATEN, J. F.: *Thermal Performance of Buildings*. Elsevier, London, 1967.
5.11 LOUDON, A. G.: *Summertime Temperatures in Buildings Without Air Conditioning*. Building Research Station Current Paper, CP 47/68, 1968.
5.12 MILBANK, N. O. and HARRINGTON-LYNN, J.: *Thermal Response and the Admittance Procedure*. Building Research Establishment Current Paper, CP 61/74, 1974.
5.13 BASSETT, C. R. and PRITCHARD, M. D.: *Heating*. Longman, London, 1969.
5.14 BURBERRY, P.: *Environment & Services*. B.T. Batsford, London, 2nd edn., 1975.
5.15 CROOME-GALE, D. J. and ROBERTS, B. M.: *Air Conditioning and Ventilation of Buildings*. Pergamon Press, Oxford, 1975.
5.16 JONES, W. P.: *Air Conditioning Engineering*. Edward Årnold, London, 2nd edn, 1973.
5.17 CHARTERED INSTITUTION OF BUILDING SERVICES: *CIBS Guide A1: Environmental Criteria for Design*. CIBS Publications, London, 1978.
5.18 CHRENKO, F. A. (Ed.): *Bedford's Basic Principles of Ventilation and Heating*. H. K. Lewis, London, 1974.
5.19 ELECTRICITY COUNCIL: *Reciprocating and Screw Refrigeration Machines Applied to Heat Recovery*. Job no. 9/748/1, London, Oct. 1977.
5.20 AMBROSE, E. R.: The heat pump: the performance factor, possible improvements. *Heating, Piping, and Air Conditioning*, Vol. 46(5) (1974), pp. 77–82.
5.21 CHAPMAN, P.: *Fuel's Paradise*. Penguin Books, Harmondsworth, England, 1975.
5.22 AMERICAN SOCIETY OF HEATING, REFRIGERATION AND AIR-CONDITIONING ENGINEERS: *1975 Equipment Handbook*. ASHRAE, New York, 1975.
5.23 DUFTON, A. F.: *The Equivalent Temperature of a Room and its Measurement*. Building Research Technical Paper no. 13, 1932.
5.24 JONES, W. P.: Energy consumption by variable air volume systems, *Building Services and Environmental Engineer*, Vol. 2 (1980), pp. 16–19.
5.25 COURTIER, G. A. C. and NOON, C. L. B.: Energy conservation in the ventilation of buildings, *Building Services and Environmental Engineer*, Vol. 2 (1980), pp. 25–26.
5.26 EDWARDS, J. H. *et al.*: Humidifier fever. *Thorax*, Vol. 36(6) (1977), pp. 653–663.
5.27 DEPARTMENT OF HEALTH AND SOCIAL SECURITY: *Press Release Guidance on Legionnaires' Disease*, DHSS, London, 1980.

Chapter 6

Dynamic Performance and Energy Consumption

6.1 INTRODUCTION

Much of the discussion in Chapter 5 involved simple descriptions of the physical processes taking place during transient thermal conditions. During building design, it is important that the design team is provided with information based on more than simply an understanding of the transient physical processes, because it will be necessary to have them quantified whenever decisions relating to temperatures or energy are required.

There are many decisions which require quantification during design. For instance, one aspect of environmental control may require an analysis of the thermal performance of the building fabric so that condensation does not occur on cold surfaces causing problems of degradation of the building fabric and mould growth. A more sophisticated quantification is necessary when assessing the requirements and solutions for thermal comfort and energy control. Furthermore, the dynamic performance of the building fabric and the dynamic performance of any mechanical services must be adequately matched to ensure thermal control of sufficient accuracy for the internal requirements. Incompatibility between the building services and the building fabric can result in unnecessary capital expenditure and unnecessarily high running costs, and so a detailed understanding is important for making decisions.

In any country or climate, the design of the building must aim to achieve the relevant objectives from the brief, whether implied or explicitly stated, and to meet design standards imposed by legislative building controls. For instance, the 1978 amendments to the Building Regulations for England and Wales [6.1] were aimed at encouraging more consideration of the energy consumption of buildings whilst attempting to retain design flexibility (Part FF—*Conservation of Fuel and Power in Buildings other than Dwellings*). Designers who choose to demonstrate the success of their design for energy conservation are given the opportunity to do so by applying for approval under Section FF 3, which does not stipulate a performance standard but merely demands:

> 'adequate resistance to the passage of heat, the loss of which . . . would entail the consumption of fuel or power to enable temperature conditions normal for the proposed use of the building or part to be maintained.'

The regulation applies to the building fabric and does not specifically mention building services.

'Adequate' is left for the designer and the local authority to agree on, although an assessment of adequacy may be obtained by comparing the proposed building with a notional building which satisfies the relevant deemed-to-satisfy provisions of the regulations [6.2].

There are several approaches to executing the comparison, including the degree day method, energy targets [6.3], calculator packages and computer based techniques [6.4]. Examples will be discussed in more detail later but it is important that the method of comparison should include facilities for the incorporation of weather data, the thermal performance of the building and the thermal performance of the building services. Hence, legislation encourages the use of prediction techniques which cover these aspects.

The building regulations are intended to operate as a minimum standard and solutions to higher standards will still receive approval. The overall standards which a proposed building is to achieve require consideration of building costs, energy costs, maintenance costs, expected life and changing values of energy throughout the expected life. Hence, a given brief would result in buildings of different standards in different countries or climates.

Deciding how to meet or improve upon the standards requires input from many members of the design team during the design process. They must decide upon the level of investment instead of merely designing to the minimum standards imposed by legislation. The task is then to analyse the economic benefits of reduced operating costs and improved environmental conditions against increased capital costs. (It is assumed that capital costs will always increase with improvements, otherwise if an option was better and cheaper it would already have been incorporated into the scheme.)

These operating costs are not only the energy requirements for thermal control. The analysis requires consideration of all the significant operating costs and how they are affected by the building design. For example, window cleaning costs, artificial lighting costs and office cleaning costs are all influenced by daylighting requirements, plan form, area of glazing, type of glazing, shading arrangements, intended plant and furniture layout and solar gain. Hence, this optimisation process should be a total optimisation incorporating all features which significantly affect the operating costs. This is a complex task requiring experienced building design team members, and only one part of the task is to predict the operating costs caused by the consumption of energy. It is worth noting that energy costs are generally one of the more significant operating costs and so it is prudent that some effort is always put into the consideration of energy control.

6.2 TRADITIONAL METHODS FOR PREDICTING ENERGY CONSUMPTION

6.2.1 Rules of Thumb

Many rules of thumb have been developed from experience of heating costs. They are usually based on the boiler load or a typical energy consumption per unit volume of the building. Their accuracy is not quantifiable but is related more to the experience and intuition of the user.

Such techniques are insensitive to the details of the architecture or services and so they are not applicable for in-depth analysis of energy consumption.

6.2.2 Degree Day Calculations

The energy consumed by buildings for winter heating relates to the duration and severity of the weather. Provided the design steady state heating load is known the energy requirements should be predictable using the degree day technique described in Section 3.4.

The degree day technique should be most successful at predicting energy requirements for continuously heated buildings as the application is based on a simplified representation of the actual process of heat transfer. Unfortunately, accuracy is not particularly good [6.5] and it is further affected when using correction factors to allow for reduced energy consumption when heating services are operated intermittently [6.6]. It has been estimated that the calculated energy requirement could be within $\pm 25\%$ of the actual energy requirement [6.7].

Rising fuel prices have encouraged a move towards lower energy consumption and this is often achieved by designing the building to a lower base temperature (Section 3.4) by, for example, a generous use of insulation in appropriate locations, sensible use of glazing and minimal external surface areas. However, degree days are currently published in the UK for base temperatures of 15·5 °C (standard degree days) and 18·5 °C, whereas many buildings are designed to much lower base temperatures. Factors are published to convert standard degree days to degree days at base temperatures between 18 °C and 10 °C [6.6]. However, more factors are now required for 'low energy buildings' which are being designed to base temperatures around 5 °C [6.7].

Although the degree day technique is a quick and simple method of indicating the approximate annual energy consumption for heating, it is of little assistance when detailed comparison of different design options is required. The inherent simplicity of the calculation renders it unable to cope adequately with the finer details required in the majority of such design analyses.

The technique is not applicable to the prediction of energy consumption in cooled buildings because it is insensitive to the features which significantly affect energy consumption, for example, solar radiation heat gain, relative humidity and cooling services design.

However, it does have a use as a quick and simple technique for obtaining a guide to the energy requirements for heating a proposed building or for an existing building where energy control measures are being proposed, provided the results are used in a manner which reflects their inherent accuracy. The technique is suitable for operation on programmable hand held calculators which can also draw upon additional information from a fixed data memory store. Casual heat gains can be included in the calculations but the more complex solar radiation calculations still remain a difficulty for small programmable calculators and this is sometimes overcome by a statistical allowance related to window area. A different technique uses averaged annual solar radiation data with tabulated numbers of days at a variety of averaged temperatures [6.40].

6.2.3 Electrical Analogue Simulation

Similarities between basic heat flow equations and basic electrical equations enable transient temperatures to be simulated by electrical circuitry. The electrical circuit is designed so that the thermal properties of the building materials are represented by electrical resistors and capacitors for each element of the building (walls, floors, roof, glazing). These electrical circuits are then assembled with other circuits which represent the air within rooms and external air, to form a complete model of a building (e.g. [6.8]).

The heat flow into the building is created by injecting an electrical current into the relevant point in the electrical network (e.g. convectors—heat into the air within rooms; solar radiation—heat into the external surface of external walls and penetrating windows to reach the internal surface of floors or walls). The voltages which arise throughout the electrical network are related to the temperatures which will occur in the actual building and these can be plotted directly onto temperature versus time graphs.

The simulation can operate on time scales which speed up the predictions. For example, one hour in reality can be simulated by a few seconds in the model so that large amounts of data can be generated rapidly. Hence a year's simulation can be undertaken quite easily, to provide annual energy consumption data (e.g. [6.9]).

This type of modelling is performed on an 'analogue computer'. For applications in design work, the computer circuitry must be changed whenever the design team wishes to change the architecture or construction. This is not always achieved easily on simple analogue circuits and there can also be quite complex problems generated by changing weather and solar radiation. Although analogue computers have been developed to quite a high degree of sophistication, these problems, combined with restricted accessibility to analogue computers in comparison to digital computers, have contributed to the increasing preference for digital computers for temperature and energy predictions.

6.3 DEVELOPMENTS IN PREDICTION TECHNIQUES

Progress with the prediction of energy consumption has occurred through methods which simulate or model the transient thermal environment in and around buildings using digital computers. There is a wide variety of different techniques which have been developed to provide solutions to the relatively complex equations expressing changing temperatures as time progresses. These techniques have been incorporated into many computerised models which simulate the thermal performance of buildings [6.41]. It is not necessary to understand the detailed operation of these techniques but the generalities are relevant [6.10].

6.3.1 Generalised Operational Characteristics of Computer Simulation

Most of the computer techniques operate by calculating the change in temperature at discrete time intervals for every element of the building structure. Each element is simulated individually and the thermal interrelationships with other elements, such as convection between air and surfaces and radiation exchange between surfaces, are taken into account in the calculations at each time interval.

There is scope with high speed digital computers for the simulation of a large amount of detail. This is particularly convenient when the detail is already available as output from computers, such as weather data tapes for a typical year. Only the relevant information need be extracted from the data tapes by the computer for use in the calculations.

The task is not so convenient when data have to be obtained manually from drawings or design decisions and entered into the computer. Often these data are required in two forms. Firstly, there are the data describing the construction materials and how these materials are built up to form the elements of the building. These data may need to be changed as the building design progresses or when a decision is required regarding an optional form of construction. Secondly, there are the data describing the activities which will take place inside each room of the building as time progresses in the computer simulation.

Both types of data can be tedious to prepare and so the selection of a computer technique requires consideration of the amount of manual work necessary to input the data into the computer, the ease with which changes can be made and also some indication of errors in the results as well as possible errors in the input data. However, it is not always necessary to obtain a computer simulation with the minimum of error because many design decisions only require an indication of the value of a modification. This provides the opportunity for a wide variety of short cuts to be introduced into the mathematical modelling, not only to reduce the amount of work the computer has to perform, but also to reduce the amount of data which must be supplied manually. For example, an air conditioned office building might be simulated by taking only one typical room of the multi-roomed offices when analysing the reduction in energy consumption caused by, for instance, internal blinds, solar control glazing or external solar shading devices.

The increase in error with the introduction of short cuts is not usually quantified. In fact, it is often difficult to obtain any indication of the errors with some computer techniques. Predictions using digital computers should be able to achieve high accuracy and flexibility, but there are many factors which influence this. For example, there is a wide choice of approximate numerical solutions to the complex heat transfer equations, each with differing requirements of computing effort and producing results of varying error. Commercially, there is little benefit in using a solution with the minimum of error because:

(a) The computing effort may be considerable, requiring long and costly calculation time.

(b) The computer requirements for the storage of large programs and data may become expensive.
(c) The error in the technical data on the physical properties of building materials and heat transfer coefficients becomes more significant as other errors are reduced.
(d) The design team often have only a general impression of how the building will be used.
(e) Subsequent changes during design may render any one study obsolete.
(f) The computer is only an aid to design, usually assisting by comparing one design with another.

Nevertheless, computer simulations should be inherently more accurate than the approximate methods developed for hand calculations because they operate in a more complex manner, and they also have the potential for more detailed investigations.

When digital computing techniques were first introduced, different computer programs were compared with each other for the same simulation exercise and this generated a wide variety of different answers to the same problem (e.g. [6.11])! This is not unexpected because, apart from the different mathematical approaches, computer programmers will have their own particular method of dealing with a given problem. These computer programs involve a large number of problems which do not have standard solutions, leading to a variety of different programming techniques. On the technical side, there is still a dearth of information regarding certain heat transfer processes although theoretical and practical research has provided some of these data (e.g. [6.12, 6.13]).

A different exercise compared measurements in a room thermally monitored with instruments with the results from a computer simulation technique developed for research purposes [6.14]. This attempted to predict the conditions in the room knowing only the construction of the room, the weather conditions, heat inputs, ventilation and other relevant factors. The results of this research and from other research projects, and the data now accumulated from real applications of computer aided design studies, have indicated that the errors can be expected to range from insignificant to 5%, and this would be considered adequate for the majority of design decisions [6.15].

However, other comparisons of computer programs for energy analysis [6.42] have shown sufficient disagreement to undermine confidence about some of the programs, particularly for cooling applications, without undertaking elaborate, comprehensive and expensive research projects to compare the predictions with reality [6.16].

The introduction of short cuts to reduce computing effort and manual data input are usually for commercial reasons. Some computing techniques are merely the approximate methods used for hand calculations and obviously cannot achieve a result any more accurate than when performed manually, although there can be considerable time savings when multiple calculations are required.

When selecting a method of computer simulation, fundamental details of the computing techniques should be obtained to ascertain whether the technique is one of the hand calculation techniques which has been programmed into the computer or one of the numerical methods specifically developed for solutions to the complex heat transfer equations by digital computers. Some designs use the somewhat misleading term 'computer aided design' even when the computer is only performing calculations originally developed for manual use!

More comprehensive discussion regarding the evaluation of computer programs can be obtained elsewhere [6.17].

6.3.2 Simulating the Performance of the Building Services

If it is decided to simulate the thermal performance of a building by computer, the computer output can also provide the basis for the design of the mechanical services for heating, cooling or ventilating the building.

By simulating the performance of the services, it is also possible to compare the effectiveness of different services designs at providing the required environmental conditions. For example, a study may investigate the benefits of a cooling system for summer conditions when over-heating is predicted to occur. Conversely, the study may examine different methods of preventing summertime over-heating without cooling plant by using, for instance, increased ventilation or solar control devices.

When general conclusions have been reached on the active environmental control systems necessary, various options can then be analysed in detail for their suitability and effect on energy consumption. For example, if a building with a quick thermal response requires a heating installation, the analysis might examine a quick response warm air system compared to a relatively slow response wet heating installation (e.g. radiators). The interaction between the dynamic performance of the building and the dynamic performance of the heating installation can then be simulated by the computer to obtain the best decision in terms of cost and energy consumption.

6.3.3 Other Benefits of Computer Aided Design

Although the prediction of overall energy consumption is often the principal motive for using computer simulations, these techniques can be used to generate a considerable amount of other information for the building design team. For example, details of calculated temperatures for constructional elements of the building may be of interest for condensation control or thermal comfort at critical conditions during the year. Similarly, the energy transfer across specific elements such as windows, walls and roofs may be the subject of particular interest when considering constructional options for an individual room. When a computer is modelling a building in detail, all of this information can become readily available to the design team so that the eventual design of the building is improved. In fact, once a building is 'on the computer' there is then the possibility of obtaining information in other areas such as natural and artificial lighting design, budget costings and bills of quantities for contractors to price. Although such computing systems are possible, no complete system is known to be commercially available at present. However, with thermal aspects, there are many studies which are being researched and some of the conclusions are presented in the following section as examples of the application of computers to help improve building design.

6.4 CONSIDERATION OF SOME ARCHITECTURAL DECISIONS

Many architectural decisions affect energy consumption and it is not proposed to attempt to cover them all. Instead, it is intended to illustrate how computers can aid design by quantifying the benefits of different solutions.

Descriptions of basic approaches for reducing energy consumption in building design are available from a variety of sources (e.g. [6.9, 6.18, 6.19, 6.20, 6.21]); these deal with generalities such as external surface area, plan, form, layout and materials selection, while examples of case studies are given in [6.22, 6.23, 6.24]. Computing techniques can be used to investigate the effect of modifications to each of these aspects and probably the most appropriate general investigation is to consider the optimum quantity and location of thermal insulation. The following examples are presented to illustrate the variety of analyses which can be examined and they are not intended to indicate priorities.

6.4.1 Orientation

Rotating a building on an unrestricted site can influence the energy requirements. A major cause is the changing amount of solar energy captured by glazing for each room of the building.

In hot climates, the amount of energy collected should be controlled to prevent excessively high temperatures or unnecessary loading of cooling plant. At the other extreme, buildings requiring heating for most of the year can be designed to maximise capture of solar energy by the glazing. Between these two climatic extremes the objectives change with the time of year. This may lead to mechanical systems designed to vary the solar energy collection depending on, for instance, the particular requirements of a room.

The detailed analysis of such options is complex but can be achieved by computer simulation. Results have been obtained from a computer study which investigated the orientation of terraced houses on an estate using meteorological data for Manchester, England [6.14]. The houses were separated by the minimum distance for the recommended daylighting and sunlighting, but there was still some shading of parts of the houses by other houses on the estate when the sun was low in the sky in winter.

The computations simulated a family of four moving around the house and engaged in typical activities in each room in the dwelling at relevant times. In each room, a simulation was made of a thermostatically controlled electric convector heater, with an instant response and 100% efficiency, which controlled temperatures between 0700 h and 0845 h and between 1530 h and 2250 h. This was intended to represent both parents working and the children at school, leaving the house completely unoccupied during the middle of the day. Ventilation was related to the number of occupants in each room at any particular time and allowances were made for uncontrolled heat gains (e.g. television, cooking, refrigeration, lights and people), and for the thermal capacity of furniture. Heating thermostats were set to the following air temperatures:

Room	Temperature (°C)	Location in the house
Lounge	23	Ground floor front
Dining room	21	Ground floor rear
Hall	18	Ground floor front
Kitchen	18	Ground floor rear
Bedrooms (3)	20	First floor front and rear
Bathroom	20	First floor front

The energy consumed by the heating installation during each time interval in the computations was summed and the annual consumption obtained for four different orientations, north, south, east and west.

The results indicated that the lowest energy consumption occurred with the houses facing due south, irrespective of whether the houses had heavy or lightweight internal partitions. In the lightweight houses, east facing houses required 5 % more energy than south facing houses and north and west facing houses required 2 % more than south facing houses. Similarly, in heavyweight houses, east facing houses required 5 % more energy and west facing houses required 4 % more energy than south facing houses. For comparison, the energy lost by the ventilation air amounted to around 50 % of the annual energy consumption.

These results cannot be regarded as generalisations because of architectural features peculiar to this project. However, the quantified variations in energy consumption enable the design team to make judgements based on the actual reductions in annual operating costs.

6.4.2 Thermal Response

There is often a common misconception that intermittent heating or cooling does not save energy compared to continuous heating. Consideration of the heat loss from, say, a domestic hot water cylinder heated by an electric immersion heater may clarify this. Sections 1.1 to 1.5 showed that heat loss is a function of differences in temperature. Hence, the longer the hot water is held at a high temperature (i.e. the temperature difference is a high value) the greater will be the heat loss and energy consumption. The only proviso is that consideration must also be given to the performance of the services which may operate inefficiently under frequent on/off cycling but this is not the case with an electric immersion heater. However, the loss of efficiency during intermittent plant operation can be reduced by appropriate decisions by the services engineer during plant design. Hence, the objective is to hold high differences in temperature across the building fabric for the minimum duration and so the more rapidly a building can be brought to the required temperature, the less the energy requirement of the building.

It is clear from Section 5.2 that the thermal performance of a building is affected by the amount of material within the building and the manner in which that material absorbs heat during a change in temperature. As more material is located inside the building, so the thermal weight of the building is increased. The location of thermal insulation in the construction of the envelope can also affect the thermal mass of the building, as well as influencing the amount of heat conducted through the fabric. If the insulation is located on the inside face of walls, roof and floor, this partially isolates the thermal mass of the construction from any change in the internal temperatures. By increasing the thickness of the insulation, the amount of isolation increases. Hence, any heating of the enclosed spaces will have little effect on the temperature of the envelope itself and so more of the heat will be usefully raising the internal air temperature and the temperature of partitions, intermediate floors and any materials in storage. Because the envelope is not being heated so much, the same amount of heat will cause temperatures to rise more rapidly. The same effects will occur with cooling systems, i.e. temperatures will fall more rapidly for the same amount of heat extraction.

Conversely, if the insulation is placed on the outside of the envelope, the thermal mass of the envelope will be involved in the changes in internal temperatures.

With some lightweight buildings, the actual location of the insulation in the envelope will not influence the thermal response significantly because there is negligible thermal mass internally or in the envelope. The location of the insulation in the envelope of buildings of masonry or concrete construction can have a significant influence on the thermal performance of the building when the building has a negligible thermal mass internally (e.g. timber suspended floors and lightweight partitions) but this effect is lost when there is a considerable mass within the building due to the internal structure (e.g. concrete floors or walls) and heavy materials in storage (e.g. warehouses or furniture).

Consideration must be given to the location of the insulation in the construction when the thermal response of a building is being designed to influence the energy consumption. This only applies to buildings in which temperatures are controlled intermittently. When buildings are held at a constant temperature the thermal response of the building is of little interest and so the insulation can be designed for other criteria.

An extension of the housing study by computer simulation [6.14], described in Section 6.4.1, compared lightweight and heavyweight dwellings, all of which faced south and contained the same amount of furniture. The lightweight dwellings had lightweight partitions and intermediate floor, and their external cavity brick walls were insulated internally; the heavyweight dwellings had solid partitions and their external cavity brick walls were insulated either in the cavity or on the outside.

The study concluded that all the heavyweight dwellings required more energy. Table 6.1 shows how much more energy was required for the

Table 6.1 The extra energy required by heavyweight dwellings compared to lightweight versions of the same dwelling (computer simulation)

Description	Extra energy required by heavyweight dwelling
Uninsulated dwellings (*U*-values 1·86 W/m^2 °C)	
Approximately 50 % wall area glazed	12 %
Minimum glazed areas	14 %
Insulated dwellings (*U*-values 0·56 W/m^2 °C)	
Approximately 50 % wall area glazed	18 %
Minimum glazed areas	19 %

heavyweight versions of each option; the range was from 12 % to 19 % more energy than the lightweight versions. Although no tests were conducted, the *differences* in energy consumption could be expected to decrease if the dwellings had temperatures controlled for just one period per day (e.g. all day or all night) and to decrease further if the dwellings were continuously temperature controlled. Actual energy consumptions would of course be expected to increase the longer the controlled temperatures are maintained.

The reasons for the heavyweight dwellings requiring more energy could be obtained from a room by room analysis of the computer output, one sample of which is shown in Fig. 6.1. The energy going into storage in the structure of the dwellings was being released when the heating was turned off. Hence, as more energy was stored in the structure of the heavyweight dwellings, they were held at higher temperatures when the heating was off. Consequently, the difference in temperature between inside and outside was greater and so the energy transfer by the fabric and by ventilation was greater than for the lightweight dwellings. In fact, the absolute increase in energy consumption in each comparison was directly related to the mass which was heated and cooled during each cycle. The advantages gained with lightweight buildings must be compared with any disadvantages ensuing. Any penetration of solar radiation when conditions are already comfortable without heating could cause rapid over-heating. Hence, consideration should be given to solar control whenever lightweight designs are likely to cause over-heating discomfort. The same effect can be seen in buildings cooled to remove solar radiation heat gains. The cooling plant must be designed to remove the peak loads if the structure is limited in its ability to absorb heat, whereas a heat absorbing structure will take some of the load off the cooling plant provided slight temperature variations can be accepted. The design features which influence thermal response are

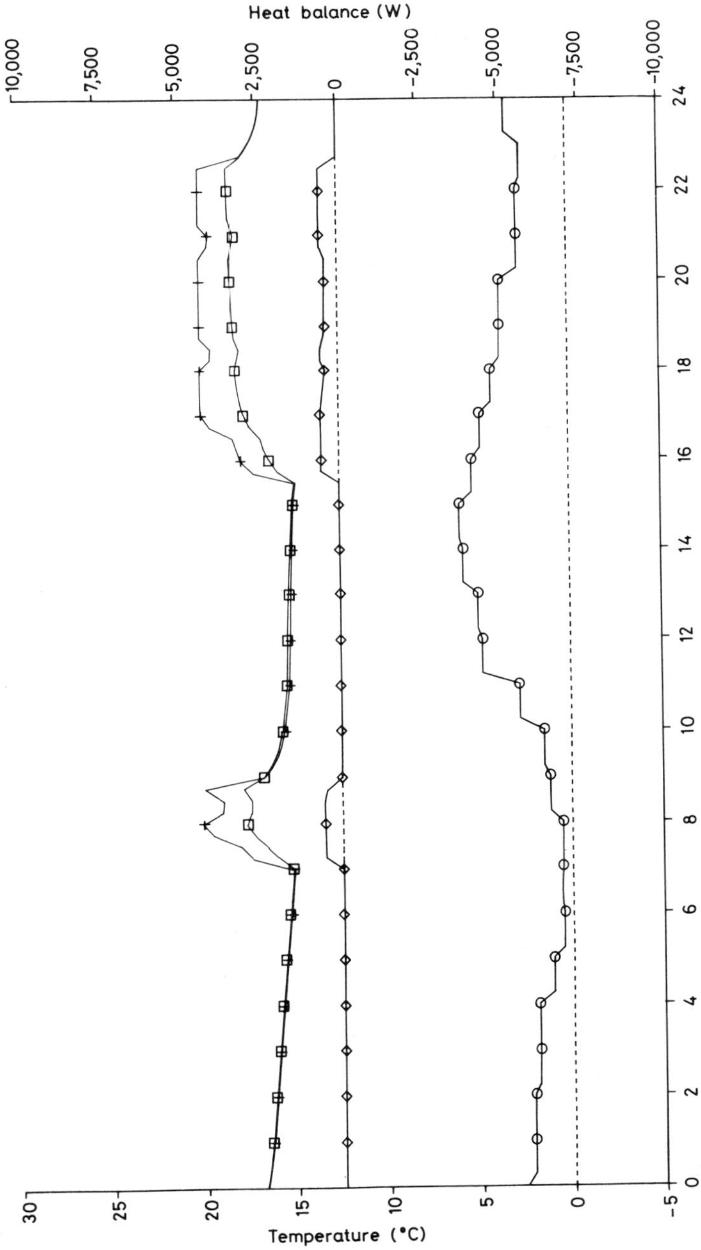

Fig. 6.1. Typical example of computer graphics, showing the simulated temperature conditions in a particular room of a building over a 24-hour period, the heat output of the heating installation in the room (heat balance), both of which were computed for the (recorded) outside temperatures (also plotted) (from [6.14]). ○, Outside air temperature; +, air temperature in room; □, mean radiant temperature in room; ◇, heat balance of the room.

particular properties of the construction materials. For short term changes in temperature (e.g. less than one hour) only the properties of the surfaces to a depth of around 25 mm need be considered. Long term changes (diurnal changes) require consideration of all the materials in the construction. The relevant thermal property is the thermal penetration coefficient explained in Section 5.2.1. Typical values are given in Table 5.1.

Although the building designer can influence the thermal response of the building by the selection of materials of particular thermal penetration coefficients, this is not a quantitative design process until the overall effects are obtained from computer simulation. This also involves the ability of the building services to achieve the desired temperatures within a sufficiently short period of time, necessitating an installation which may have a capacity greater than the maximum steady loads the installation would have had to accommodate. Hence, computer simulation should include the performance of the services as well as the building when analysing thermal response.

6.4.3 Glazed Areas and Orientation

In climates where heating is required, buildings have been designed with the objective of collecting solar energy by the 'greenhouse' effect (Section 1.3.6) (e.g. [6.25]). However, whilst energy is being collected through glazing it is also being lost principally by conduction and convection from the glass to the outside air and this must also be taken into account. As the difference between inside and outside air temperatures increases, so the heat losses increase whilst the solar radiation heat gains remain the same. This could result in the heat losses exceeding the gains, particularly if the internal temperatures are controlled to high values or the external temperature is low, and so the heat energy requirement for the building would be reduced if the glazing was replaced by an insulated wall construction.

An analysis of this situation requires data on the local meteorological conditions, the periods for which the particular temperatures must be held and the orientation and shading details of the glazing. The heating services must be designed to reduce heat output when solar radiation is contributing to the room requirements, otherwise overheating will occur and the energy benefits will not be obtained.

No generalised solutions to this problem are available yet, although theoretical and practical studies have been published for a few situations. The following are typical examples.

Regarding orientation, a dynamic simulation model [6.26] has been used in a study [6.27] to analyse the energy consumption and thermal behaviour of a solar house in Israel (latitude 32° North, longitude 35°). The annual heat energy consumption was predicted for the house with either a window in the south wall or in the north wall of

(a) an uninsulated building,
(b) an insulated building.

The results confirmed that less energy was consumed when the window was in the south wall for both uninsulated and insulated houses, and the actual saving was quantified at approximately the same in each case.

A consideration of the area of single glazing was made in the study of terraced housing at Manchester (see Section 6.4.1). It was concluded that the minimum area of glazing in the north walls caused the least consumption of heat energy but this was not so important for the irradiated windows of the south wall, particularly for the lower temperature rooms. In fact, some south facing rooms consumed around 10 % more energy when the glazing was reduced from 50 % of the external wall area to the minimum for adequate daylighting. From this study it was concluded that the energy benefits of different glazed areas should be studied on a room by room basis unless rooms and their temperatures are identical. Another study [6.28] concluded that east and west windows are also net contributors to the energy requirements of buildings at Bracknell, England (latitude 51·4° N) provided they are double glazed.

6.4.4 The Degree of Exposure of Glazing

The degree of exposure of the whole building is a general consideration in design and it may be considered worthwhile to influence the local environment by, for example, planting trees around the building to reduce the influence of wind and clear skies on the external surface heat transfer. Because glazing can have significant effects on the thermal performance of buildings, its exposure is often given more detailed consideration, the analysis of which could be assisted by computer studies.

The location and angle of the glazing in vertical window apertures affects the energy consumption. Glazing placed flush with the external surface of the building envelope or even projecting beyond the surface will be exposed to the severity of the weather. Air movement across the building facade will impinge on the glazing, causing increased convective heat transfer and greater cooling or heating requirements. Projecting glazing will in general have reduced shading from direct sunshine unless the glazing is angled so that most of the solar energy is reflected or there are other external shading arrangements. The surface of the glazing will receive more wetting during rain, sleet or snow, and this causes more conductive and evaporative heat loss. Overnight, the glazing will radiate more heat to the sky which behaves as though it is a surface, the apparent temperature of which can fall to around $-40\,°C$ during clear sky conditions in temperate climates. Hence, considerable radiation heat loss can occur from glazing which can 'see' the night sky.

Preventing such energy transfer is simply a matter of protecting the glazing from wind, sun, rain and excessive radiation losses. The simplest solution is to recess the glazing from the external facade. The effect can be enhanced by horizontal shading features such as projecting eaves, balconies and louvres, and vertical shading features such as projecting columns, remote structure and louvres. Landscaping can also assist by providing sheltering using mounds or vegetation. One of the useful landscape features in temperate climates is the use of deciduous trees for solar control, whereby shading is provided in summer but when the leaves have fallen in winter solar energy can penetrate into the building to provide useful heating.

Such features must be designed with care to avoid undesirable influences. For instance, projecting concrete floor slabs which provide useful shading of windows can also act as 'cold bridges', creating heat transfer fins on the external surface and causing greater convection loss and overnight radiation loss. Internally, mould growth may occur on the cold surfaces. Another example is deciduous trees situated adjacent to buildings for shading; their roots must not undermine foundations, their leaves should not block surface water drainage systems and any energy benefits may be financially worthless if the cost of a felling resulting from such problems exceeds any savings.

Whilst it would be most desirable to analyse by computer simulation the energy benefits of the degree of protection of glazing, this is one area where the shortage of technical data impedes progress. Hence, building design must progress in the meantime without such assistance and instead design decisions are based on the general application of the basic principles of energy conservation. The quantification of actual savings by computer simulation, whilst technically possible, cannot be accomplished with complete confidence as the data relating meteorological conditions around buildings, glazing location and degree of exposure are not known comprehensively. For example, whilst the relationship between convection heat transfer and surface air velocity is known (see Section 1.4.2) and wind measurements for many sites are known (e.g. [6.29]), the relationship between the wind speed and direction, local topography, building shape and the exposure of the glazing cannot be interpreted into an air velocity across the glazing with sufficient accuracy [6.42].

6.4.5 Services Layout

Service networks should be kept to a minimum so that energy wastage in distribution is minimised. As this will minimise the thermal mass in the services system there will also be reduced energy wastage if the services operate intermittently (see Section 6.5.1). Hence, the plant installation and the locations which require its services should be kept as close to each other as possible. However, there will be other constraints imposed on the centralisation of plant by the overall design of the building and so a compromise is necessary as illustrated by the following example.

Consider the siting of heating plant serving a large establishment. The heating plant should be located roughly central to the loads it must serve. This will minimise heat loss from the insulated heating distribution mains, and minimise the water content and metal content of the heating mains, which affects the energy dissipated from the mains when they cool during periods of shut-down. However, this must be compromised by consideration of the following:

(*a*) *Fuel supply*. Road or rail access is required for solid fuel deliveries to the storage location. A pipeline to convenient road or rail locations for liquid fuel deliveries must be provided. Gaseous fuel can usually be piped in without major problems.

(*b*) *Fuel storage* (*where necessary*). Space is required for storage of sufficient fuel to maintain services in operation between programmed deliveries or during discontinuity of supply.

(*c*) *Fresh air requirement*. Combustion requires air, and it is considered better design to supply air naturally instead of installing more mechanical services to duct air to combustion equipment. Additional ventilation may be required in the plant room for temperature control.

(*d*) *Flues*. The combustion products must be vented to atmosphere, and may need to be cleaned or scrubbed to control pollution. The most appropriate location must take account of the locations where fresh air is required for input to the building and the prevailing winds and wind patterns around the building. Otherwise combustion products may enter the building through open windows or ventilation apertures.

(*e*) *External aesthetics*. Fuel storage, combustion plant, distribution mains and the flues from combustion equipment can be quite dominating and it is important to consider the interaction of such features with the remainder of the building form.

(*f*) *Maintenance*. The cleaning, servicing and replacement of combustion equipment and associated services all tend to be dirty operations and direct access from the outside is usually preferable to entry from the building interior.

(*g*) *Plant life*. The plant usually has a life which is much less than the building's life and so provision must be made for some heavy items of equipment to be removed and replaced, perhaps two or three times during the building's life.

(*h*) *Noise and vibration*. The noise and vibration from the plant can either be controlled by additional expense on control measures or by locating the plant where any noise or vibration does not cause disturbance to the building occupants or the neighbours. (This assumes that the noise or vibration is an inherent problem and is not caused by inadequate design or installation.)

(*i*) *Distribution direction*. If the distribution mains on leaving the plant room have to be routed outside the building, the heat is lost to atmosphere, whereas heat lost from internal distribution mains may assist the building's heat requirements.

A list of constraints could also be compiled for other mechanical services plant, and some constraints are essential for successful operation, for instance the orientation and positioning of solar collectors if these are included in the services design.

6.5 CONSIDERATION OF SOME BUILDING SERVICES' DECISIONS

The efficient use of energy for temperature control by building services has been introduced in Section 5.3. The following is complementary to the previous discussions and considers the dynamic behaviour of services and controls, and resulting implications. Guidance for designers is also available elsewhere, e.g. [6.19, 6.30, 6.31].

6.5.1 Thermal Response of Heating and Cooling Equipment

There is often misunderstanding about the response of equipment to demands suddenly placed on it. There is usually a preference for a system to respond instantly to a change, however small. Whilst such a performance is often welcomed, the effect is sometimes lost by influences which are external to the system.

The thermal response of a building to a sudden change has already been discussed in general terms (Section 5.2) and it is clear that there can be a substantial damping effect by the thermal mass of the structure or heavy materials in storage. Hence, in such circumstances there is little advantage to be gained by having a mechanical system with a very fast response. However, a thermally lightweight building which responds rapidly to a change must have a system with a similar or faster rate of response in order to achieve an acceptable degree of control of the internal conditions.

The response of a mechanical system controlling temperature or humidity is often a complex series of events, each with their individual responses. Consideration must be given to each event and its response, and also the time taken for the output of the system to affect the conditions within the space. A few examples will serve to illustrate these points.

Consider a simple conventional electric radiant heater using a high temperature heating element and reflector to direct radiation onto a person. A series of events occurs. A person decides to turn on the heater. When the switch is made, electricity instantly flows through the wire and its temperature rises rapidly. It soon begins to glow a dull red and heat is radiated rather weakly. The temperature continues rising at an ever decreasing rate until the heater is said to be producing its full output under steady conditions. As heat radiation travels at the speed of light, there is no delay in the transfer of heat from the hot element to the person, who therefore feels the effect immediately.

In order to quantify this response, it is usual to quote the time taken to achieve 63% of the steady state output, which is known as the time constant. Typically, an electric radiant fire would have a time constant of around one minute.

For an electric convector heater, the time constant of the heater alone may also be only one minute, but now time must be allowed for the hot convection currents to rise from the heater to high level in the room, entraining room air on the way. Some air movement will occur in the room and eventually the room will fill with warmer air down to the level of the occupant who will then feel the effect of the heater. This process may take an extra five minutes to occur in a small room or very much longer for large rooms and spaces which will absorb the heat easily.

With an electric fan heater causing air movement in the occupied zone, the heated air will be stirred with the room air so that the occupant will notice the effect of turning on the heater more rapidly.

All these examples have been based on relatively quick response electric heating systems. If the heater elements are buried in bricks or concrete (e.g.

under-floor heating or 'storage' heaters) then this may add many hours to the time constant. In fact, this delay is used to 'store' heat created from reduced rate off-peak (night) electricity so that the heat arrives in the space during the day.

When a wet heating system is in use with convectors, for example, the time constant of the whole system is lengthened by the thermal mass of the boiler, the distribution system and the quantity of water which the system must heat. Traditional heating systems, such as cast iron boilers which feed heat to column radiators with large volumes, using gravity circulation through large bore pipes, have time constants of around four hours. Solid fuel boilers have to be given several hours extra for the fire to be lit. Hence, it is clear that there is little benefit in turning off such heating systems completely overnight.

Modern wet heating systems have much faster response times because:

(a) There is a move towards low water content boilers which start automatically.
(b) Different materials are being applied to boiler construction which reduce their warm-up time.
(c) Pumps are used to circulate the water through smaller pipes.
(d) The smaller pipes contain less water.
(e) The design of heat emitters has developed so that they now contain less material and less water.

The designer can still influence the response of the system by the selection of particular pieces of equipment or types of system. For example, the water content of a fan assisted convector heater can be about 10% of the water content of a radiator with the same heat output. However, the benefit of such a selection may be lost by, for example, the need for long distribution pipes.

The response of a wet system in normal operation is not the same as the response during start-up. If the system is already at normal operating temperatures, the response to a request for a change depends on the control response and the mass of the heat emitter. A time constant of a few minutes is typical for such a change with convectors, fan convectors and low water content radiators, and this rate of response can be similar to the response times of warm air plant. Typical response times should be explored for each design whenever there is an interest in the system's response to a change.

Hence, it is clear that detailed consideration must be given to the dynamic behaviour of systems under heating and cooling cycles.

Having obtained an indication of the response of services at start-up or during operation, consideration can then be given to the question of whether the plant should be operated intermittently.

If a building requires continuous and accurate control, it is clear that the services must operate continuously and so the only consideration is the response of the services to changing conditions. When less accurate control is permitted and the load on the system drops below the system's maximum capacity, there is then the opportunity for turning off the entire plant for brief periods on a regular basis. For example, an air conditioning installation for an office in a temperate climate would be designed for continuous operation when external conditions are close to or exceed the summer and winter design (i.e. 'worst') conditions. But during other occasions when the external conditions do not cause full load on the services, they could be turned on and off and still maintain internal conditions close to the design requirements. Research has concluded that no serious occupancy complaints would occur if the air temperature fluctuations do not exceed certain values which vary with the rate of cycling per hour [6.32]. Provided all other factors remain constant the guide is

$$\Delta t^2 \times n < 4{\cdot}6\,^\circ\mathrm{C}^2/\mathrm{h}$$

where Δt is the amplitude of the air temperature fluctuations (°C) and n is the number of cycles per hour. Hence, if air temperatures fluctuated by 1 °C, then the frequency should not exceed 4·6 per hour. Increasing the amplitude to 1·5 °C reduces the permitted frequency to 2·0 per hour. In fact, in comfort situations, there is some speculation that minor deviations in the internal conditions create a more stimulating and invigorating environment [6.33] but this is not substantiated.

To save the maximum amount of energy, there will also be incentives to turn off air handling units and so some of the other factors which were assumed to remain constant may change. Reduced air velocities will influence comfort sensations and there may be other reactions such as that caused by changing noise levels as plant starts and stops.

Furthermore, it will be necessary to check that the frequency at which the plant is cycled accords with the manufacturers' recommendations. More wear is incurred by a greater frequency of starting (e.g. belt wear on fan drives) and any financial saving by reduced energy consumption may be lost by more frequent plant failure and maintenance requirements.

However, this is not to say that intermittent operation of plant is inappropriate, for it is now generally accepted that intermittent operation can save a significant amount of energy. This is particularly so when the 'off' periods become longer, and the turning off of plant overnight or over weekends by a timeswitch is now common practice when control of conditions is not required. The plant is restarted automatically to provide the desired conditions at a particular time. Usually the timeswitch is set so that the desired conditions are actually achieved in time when the external conditions are imposing the design loads, necessitating lengthy restart times.

The length of time required by plant to achieve its full duty depends on the thermal response of the plant. However, the length of time required to bring the building to the desired conditions is a combination of the thermal response of the plant, the thermal response of the building, the amount the building's temperature has deviated from the required temperature, and the amount of excess capacity the plant has over the steady state loads. This excess capacity can be increased during design to reduce the start-up time and this can result in a saving in energy due to the reduced period for which the building temperatures are held and for which the plant is in operation. Typically, heating plant is sized between 20% and 50% larger than the design steady state load [6.34] but each application must be considered on its own requirements and this justifiably may lead to plant being twice the steady state loads for certain highly intermittent applications.

Consideration must be given to the efficiency of mechanical systems which are oversized for intermittent operation. The system efficiency under the usual running conditions must be checked. The latest generation of heating plant can produce acceptable efficiency under intermittent operation and with constant running at part load conditions.

In addition to the efficiency of fuel utilisation, consideration must also be given to the life of electrical control equipment and other plant, corrosion (particularly in flues) and to the demands placed on electrical distribution networks. Cooling plant for air conditioning is rarely designed with excess capacity for intermittent operation because cooling loads are usually related to external conditions and so the plant has an opportunity to achieve the required temperatures before the peak load is faced. Furthermore, cooling plant does not usually have the good part load efficiency that is possible with heating plant and so the incentive is to design the cooling plant to just meet the peak loads with no excess capacity so that the efficiency under part load operation is maximised.

6.5.2 Control Systems

Modern micro-electronic technology is ideally suited to the control of mechanical services for the type of intermittent operation described in the previous section. Whole plant or zones can be cycled on and off to maintain adequate control of temperatures or humidity in spaces where minor fluctuations do not cause concern. The off periods can be adjusted automatically to maximise the energy saving, either during the period when controlled conditions are required or during the period when no control is required, except to protect against condensation or frost damage. The plant can then be restarted as late as possible so that the required temperature is just achieved at the appropriate time.

Hence the restart time must be varied according to the time which the services will require to bring the inside temperature to the required value; this restart time is computed from the measured inside and outside temperatures.

It is possible to apply optimal control theory to the design of the services and the services control system so that the energy consumption is minimised. This requires careful consideration of many of the factors affecting the thermal response of buildings and services, as well as the typical duration for which the plant will be turned off (e.g. [6.34, 6.35]).

The commissioning of the first generation of electronic control devices created certain problems. Such control systems require considerable information about the thermal response of the plant, the excess capacity over steady state loads and the thermal response of the building. Because this information is not readily available, these control systems may be commissioned poorly and so the potential energy savings are not fully obtained.

The second generation of electronic control systems incorporate self-learning devices so that the control system learns from its own experiences. If the plant was restarted too early, the control system 'remembers' the correction which is necessary for those particular temperatures and subsequently restarts the plant later. The corrections learnt from these experiences are used to influence the start times for different temperatures so that the device eventually learns the full range of start times under a wide variety of conditions. This overcomes the commission problems mentioned above.

The plant might also be stopped early by the same principle before the actual end of the normal controlled period. This led to such controllers being called 'optimum start and stop control systems'.

Whilst such systems may appear to be the control design engineer's panacea for energy conservation, a location must be found for the essential internal and external temperature sensors to ensure sensible functioning of the system. In a single roomed building (e.g. a warehouse such as that shown in Fig. 5.8) there is only one choice of room although detailed consideration must be given to radiation heat transfer and convection currents caused by influences other than the heating or cooling system (e.g. solar gain and night radiation loss through rooflights). Compared with multiple roomed buildings, the services and building contribute fewer different time constants to the overall thermal response and so the thermal response of the one room will be learnt by the electronic control system which should then minimise energy consumption successfully.

Multiple roomed buildings usually have rooms with a variety of orientations, a variety of thermal response characteristics, or sometimes containing a variety of services each with different thermal response characteristics (e.g. radiators in one room and fan assisted convectors in another room). So the engineer must consider which room will be selected for the internal temperature sensor for the optimum start and stop control systems. Should it be the room with the slowest overall response so that some energy is wasted by the fast response rooms achieving their temperatures too early? This should lead to minimum complaints from occupants. Or, should it be in a fast response room, thereby minimising energy consumption but accepting that there may be some complaints from occupants in other slower response rooms?

Where there are clear distinctions between rooms and services with different combined thermal response, then these may deserve separate control by zoning, particularly when temperatures must be held for different periods of time (e.g. plant serving different departments or different buildings). Each zone can then have its own internal temperature sensor and optimum start and stop control system. Hence, different zones can start and stop at different times but co-ordination with centralised services may then be required.

The external temperature sensor for optimum start and stop control systems must be located where it senses a temperature representative of the outside conditions. This would normally be sited on a wall shaded from the sun and where there is no influence on the measurement of the air temperature from factors such as exhaust air ventilators, flues from boilers, refrigerant cooling plant or heat transfer through the wall.

The electronics used in the overall controls are similar to those used in microcomputers. In fact, many services installations are controlled by centralised computer operated automation systems. In addition to energy control, the computerised building automation system can perform many other useful functions such as the metering of plant, operating durations, maintenance scheduling, the monitoring of conditions throughout the services and building, and the resetting of controls. Outside the field of mechanical and electrical services, the computer can provide useful assistance with operations such as security, fire detection and control and production control.

When plant is in operation, its output must respond to the requirements. Many heating and cooling systems have temperature sensors in the space to limit or prevent heating or cooling when the required temperature has been reached. The sensors are designed to respond specifically to the air temperature and their action is usually to control the output of heating or cooling systems. Sensors which turn equipment on and off are called thermostats and they also have a time constant which typically is no longer than a few minutes.

Consideration must be given to the action taken by a room air thermostat ('room' distinguishes this thermostat from other thermostats in the system). If, for example, the room thermostat switches the electricity to a fan assisted electric convector heater, this heating system will have a combined time constant of around five minutes which is a very fast response. Conversely, if the room thermostat controls a boiler serving a wet space heating system the time constant may be around one hour and so the time constant of the room thermostat is insignificant.

To achieve better control of space temperature the usual arrangement with wet systems is to design the system so that the thermostat controls only a small part of the installation. Typically, the thermostat might control the electricity to a fan assisted convector heater and so the fan stops when the thermostat is satisfied. For an air handling unit the thermostat might control the water flow through the coils using motorised valves so that heating or cooling is prevented when the thermostat is satisfied. When the thermostat senses that a change is necessary, and this may take many minutes, the effect occurs with the minimum of delay—perhaps another five minutes for a water filled coil.

Controlling air temperatures by on/off devices often causes rapid changes in the space temperature, leading to wastage of energy through over-heating or over-cooling. Improved control is achieved either by reducing the flow of water through coils or by varying the temperature of the water through coils. This is called modulation and is usually controlled from control devices which use air temperature sensors. In fact, one of the most accurate forms of space temperature control can be achieved by modulation of water temperature using the electronic control devices.

A similar examination of the controls is necessary for humidification and dehumidification equipment to prevent the wastage of energy by excessive humidification or excessive dehumidification.

The engineering of the control systems is a complex subject and only a few of the energy implications have been introduced here. It is important that the controls are designed for the same level of energy conservation as other disciplines in the building design process.

6.5.3 Centralised Plant or Plant Matched to Each Load?

The distribution and sizing of plant raises considerations about space allocation, maintenance, replacement, aesthetics, noise, vibration, service routes and fuel storage, all of which are important. This section draws attention to the specific consideration of dynamic performance and energy consumption.

It is an unfortunate outcome of some design processes that the smallest loads on centralised plant can be the cause of the most energy wastage. Sometimes the variety of different operating loads is not considered in full when the design engineer is sizing the centralised plant and this results in large centralised services installations having to operate to satisfy very small but nevertheless essential loads. Typical examples are domestic hot water operating from central heating services which would otherwise have closed down for the summer months, and centralised chilled water services installed for mainly process requirements during summer but having to operate in winter to satisfy a small computer requirement. It is difficult to achieve a high degree of energy efficiency in either case [6.35, 6.36].

Consideration must be given to the variety of different loads imposed on centralised plant, especially those caused by different operational characteristics or times. The part load efficiency of *all* the plant must be incorporated in any comparison of centralised plant with plant matched to individual loads. For example, the part load efficiency of centralised refrigeration plant might have to be added to the energy consumed by the cooling towers, condenser water circulation pumps and chilled water circulation pumps, and then added to the cost of providing water treatment chemicals and more frequent maintenance. This would then be compared to the energy consumed by the plant matched to the isolated load and the cost of the extra capital, separate maintenance requirements, space requirement and noise control.

6.5.4 Heat Recovery

This very relevant subject is only one of the essential considerations of services design; it is extended here to include the more comprehensive subject of prevention of energy wastage.

There are many instances where 'heat recovery' is applied with published acclaim to situations which need never have warranted additional services if correctly designed initially. A common example is 'heat recovery fans' to draw buoyant hot air from under the roof of a heated factory and to deliver the hot air to the occupied zone. Although this may successfully redistribute the warm air to the required location and reduce the total energy consumption, even after allowing for the energy consumed by the fans, the heat recovery equipment would not be necessary if the design of the services had not allowed the hot air to accumulate under the roof in the first place (see Sections 4.6 and 4.7). Heat recovery should only be considered when the design of efficient services has still left a situation warranting additional services. One example is unheated factories containing hot production equipment which causes hot buoyant convection currents to rise from the occupied zone.

Consideration of heat recovery should be extended to any situation where energy is being consumed and rejected. Can the rejected energy be recovered so that the consumption of additional energy is reduced or prevented? If so, the next question is: are the economics justified?

There are many varied examples in industrial processes where heat recovery is possible. A common example is the rejection of waste heat from air compressors. The system's performance is improved by firstly drying the air before compression, this can be accomplished by using dehumidifiers of the vapour compression or heated desiccant type, and secondly by preventing the compressed air temperature from rising. The balance of the heat rejected from these processes with any heat requirement directs attention to recovering the heat for some function specifically related to the production process. (Some heat can be re-used within the air drying process but there will still be some heat left over.) As the rate of production increases, so the amount of heat generated, recovered and available for re-use increases in sympathy.

If there is no heat requirement in the production process, the next consideration is to recover the heat from the air compressing process for *any* heat requirement close to where the air compressor is situated. Possibly the simplest method of heat recovery is to use the heat for space heating, which requires only a heat exchanger and a fan to obtain the heat recovery, although some provision should be made to reject the heat to waste if sometimes space heating is not required.

Another common example where heat recovery is an important consideration is when refrigeration compressors (Fig. 5.7) are required for process cooling, dehumidification or space cooling (air conditioning). In addition to the balancing of loads, as with heat recovered from air compression, the process of recovering heat from refrigerant compression may involve designing the refrigerant condensers to operate at higher temperatures. This increases the work required at the compressor for any given cooling load, thereby reducing the coefficient of performance and so increasing the running costs. This must be considered in the economic analysis.

The ventilation of conditioned buildings also causes a loss of energy which can usually be recovered. When energy is consumed by a building's services for conditioning air for a space, the introduction of unconditioned air from outside for ventilation purposes will increase the loads on the conditioning plant. This plant may be heating, cooling, humidifying or dehumidifying the fresh air. The quantity of fresh air must be balanced by the same quantity of exhaust air and so consideration should be given to the saving of energy by exchange of heat or moisture between the fresh air and the exhaust air. (This process is hampered when the air is exhausted through many flues or through production systems which are inaccessible or heavily contaminated.)

There are a variety of methods for exchanging heat between the fresh air and exhaust air ducts. Figure 6.2 shows a rotating heat exchanger, commonly called a 'heat wheel'. It operates by rotating slowly between the exhaust and fresh air ducts. The wheel is constructed from material of a high thermal capacity, so arranged to present as much of its surface as possible to the air as the air passes through the material within the wheel. This material absorbs heat as it rotates through the warm duct and rejects heat as it rotates into the cool duct. When the material is cooled in the cold duct below the dewpoint of the air in the warm duct, condensation will occur on the wheel and moisture will be carried across to the cool duct where it will evaporate into the airstream along with the heat. Hence, both heat and moisture have been exchanged between the two ducts, thereby reducing the load on the building services.

Evaporation and condensation of moisture can involve significant amounts of energy. When moisture exchange is particularly important, the material within the wheel can be coated with a desiccant so that additional moisture exchange occurs continuously, and not just when the temperatures cause condensation.

Rotating heat exchangers usually incorporate facilities to purge the heat exchange material of the exhaust air to minimise contamination of the incoming fresh air. Nevertheless, it is not possible to achieve complete air tightness between the two airstreams and so some air carry-over is possible. Consideration must be given to the possible effects of such carry-over;

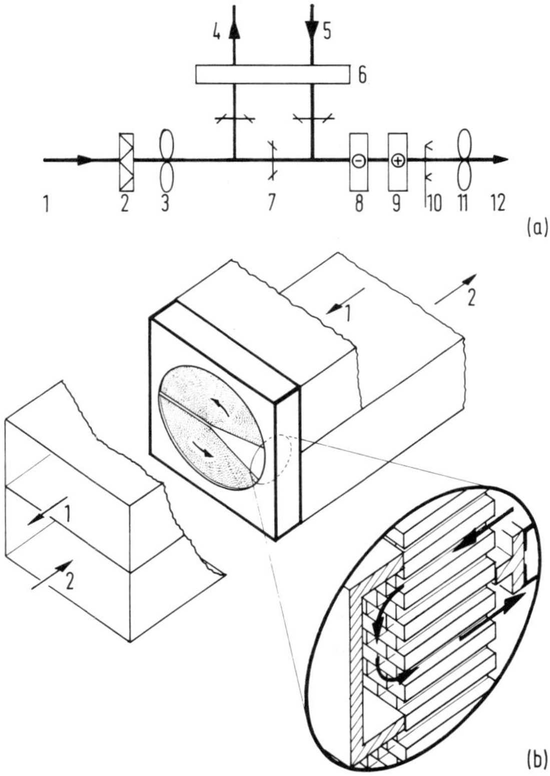

Fig. 6.2. Rotating air to air heat exchangers. (a) Positioned between the fresh air and exhaust air ducts, heat and moisture can be exchanged between the two airstreams to reduce the loads which the air conditioning plant would have had to accommodate. 1. Return air duct, 2. filter, 3. return air fan, 4. exhaust air duct (occasionally fitted with a fan), 5. fresh air duct, 6. rotating heat exchanger, 7. recirculation air damper, 8. cooling coil, 9. heating coil, 10. humidifier, 11. supply air fan and 12. supply air duct. (b) This cutaway detail illustrates a typical rotating heat exchanger installation. Where the matrix (within the wheel) rotates out of the exhaust air duct (2) and into the fresh air duct (1), a purge facility can be incorporated to clean the matrix of contaminants and exhaust air, as shown in the magnified detail.

rotating heat exchangers may not be applicable for an airstream containing a highly toxic contaminant.

A variety of heat exchangers are available which overcome the carry-over problem and also do not convey moisture. (Occasionally, moisture exchange is not desirable, such as when ventilating heated buildings for condensation control.) Figure 6.3 illustrates a few examples. Cross flow plate heat exchangers operate by passing the two airstreams on opposite sides of heat conducting plates. Condensation may occur in the cooled airstream, producing more heat for the heated airstream, but the condensate is drained away and the two airstreams remain separated. There are no moving parts, but there is no control over the heat exchange process, i.e. it cannot be turned off but must instead be by-passed if heat exchange is not required.

More complex heat transfer systems are available which avoid the need for the fresh air and exhaust ducts to be positioned adjacent to each other. Figure 6.3 illustrates the 'run-around-coil', and the heat pump is illustrated in Fig. 5.7. The 'run-around-coil' consists of a coil in each airstream, through which water (with antifreeze if necessary) is circulated by a pump.

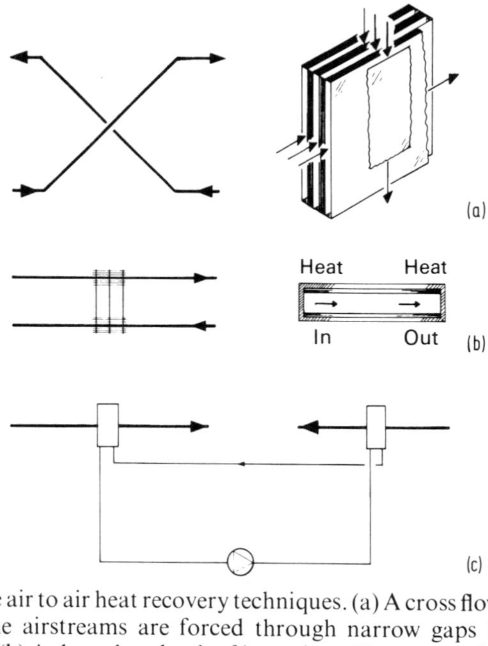

Fig. 6.3. Some air to air heat recovery techniques. (a) A cross flow heat exchanger, showing how the airstreams are forced through narrow gaps between the heat exchange plates. (b) A three deep bank of heat pipes. The section through a heat pipe shows the outer sealed casing enclosing a capillary wick. A fluid evaporates from the wick where heat is absorbed, the vapour passes down the centre of the tube and condenses where the heat is rejected. The condensate returns to the evaporating end of the heat pipe by capillarity. (c) The run-around-coil consists of a water pipework circuit, connecting coils situated in the two airstreams. The water collects heat from the warm duct and gives up the heat to the cold duct.

Heat is transferred to the water in the warm duct and is released from the water to the air in the cold duct. Condensation may occur in the warm duct if the coil is below the dewpoint and this will increase the system's performance. The heat pump is explained in Section 5.3.5.

Typically, heat exchangers transfer between 40 % and 85 % of the difference between the energy content of the two airstreams, depending on their design, except the heat pump which can transfer more than 100 % by cooling the warm airstream below the temperature of the cool airstream.

Heat exchange systems can be applied to wet systems as well as to dry systems. Frequently, industrial processes include heating and cooling of liquid products as they pass through various stages (e.g. dairy products), and this provides another opportunity for energy conservation by liquid to liquid counterflow heat exchangers between the various operations. Further details of heat exchange devices for both dry and wet applications may be obtained from [6.43].

When cooling plant is necessary for the control of over-heating, consideration should be given to the winter conditions. When the energy content of the external air falls below the energy content of the air which would have been supplied to the space from the air conditioning plant, there is then the opportunity for turning off the cooling plant and instead using external air directly. In fact, it may be worth reconsidering the design supply air energy content so that, by raising the energy content at the expense of having to handle more air, there will be savings by being able to use external air more frequently for the dissipation of energy. This facility was traditionally called 'free cooling' because only the temperature of the external air was considered, but as it is described here as a consideration of total energy content perhaps it should be called 'free enthalpy' as enthalpy means total energy content.

The demerits of additional fresh air quantities are larger fresh air plant, more control systems, increased energy consumption by fresh air and exhaust fans and increased dust loading of filtration plant, and these must be discounted from the energy benefits obtained from free enthalpy facilities. Sometimes these facilities cannot be justified economically which leads to the next consideration.

If mechanical cooling can be designed to operate at lower running costs than free enthalpy facilities, even during winter months, consideration should then be given to the use which could be obtained from the heat which will eventually be rejected by the refrigeration plant (see Fig. 5.7). This heat is not often available completely free of charge because some expenditure is necessary in order to provide the heat at the right location at the right time and at a useful temperature. Additional pipes or ducts may be necessary for getting the heat to the right location, storage for the right time and additional running costs by the refrigeration plant for the right temperature. Because of the operational characteristics of refrigeration plant (see Section 5.3.5) the rejected heat is usually at temperatures much lower than those usually employed in heating applications, necessitating larger heat emitters and possibly using forced convection principles to gain maximum heat transfer effect. Nevertheless, the cost of the recovered heat is usually a small fraction of the cost of supplying heat purchased from conventional energy sources, encouraging additional investment in such heat recovery applications due to the substantial financial savings.

The combination of motive power generation and heating applications provides one of the best opportunities for heat recovery. Whether the power is required for electrical generation, compressing air, gases or refrigerants, pumping fluids or providing transportation, the laws of thermodynamics limit the maximum theoretical efficiency of converting fuels into motive power. When losses are included, typical efficiencies are only about 30%, the majority of the remainder of the potential energy from the fuel being lost as heat. This heat could be recovered for process requirements, space heating or domestic hot water heating, providing the economics of the application are favourable. Economic considerations often limit combined heat and power projects to large scale applications in industry and to district heating applications for commercial, educational, social or domestic applications. This is perhaps one of the few occasions when the strong financial incentives to use combined heat and power exceed the incentive to match plant to individual loads.

Examples of applications of heat recovery in building design may be found, for instance, in [6.22, 6.23, 6.24, 6.37].

6.6 CONSIDERATION OF SOME INDUSTRIAL DECISIONS

The energy required for production purposes was traditionally only of significance in a relatively small number of industries (e.g. aluminium production). Changing values of all forms of energy have increased this significance to make it of importance for almost all industries, exemplified by the increasing role of the energy manager, e.g. [6.38]. Consideration of efficient energy utilisation is now very wide ranging and the following comments are intended not to be comprehensive but to highlight the issue.

6.6.1 Production Planning

Production planning should incorporate consideration of the energy consumed during each process. Figure 6.4 illustrates a typical Sankey diagram showing the sources and uses of energy for a production process. Detailed Sankey diagrams can help production directors and managers to assimilate quickly the energy flows and question whether energy discarded at certain production stages cannot be used at other production stages where heating is required. Ideally, the minimum change of energy level during the production process will give the minimum energy consumption, unless the rate of production is affected by temperature (e.g. chemical and biological processes).

Energy intensive production processes will become apparent from the

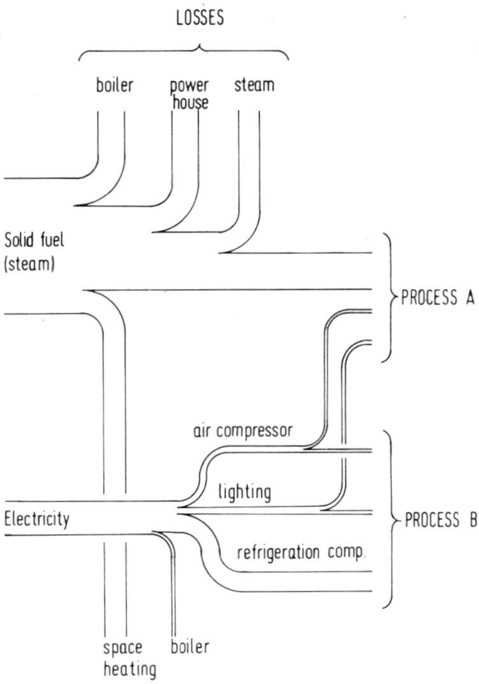

Fig. 6.4. This Sankey diagram illustrates a typical energy flow network. The width of the flow paths relates to the energy flow and produces a simple and clear method for presenting technical data to those who may not be familiar with the details of the operation. The Sankey diagram can be applied to any energy flow network and can incorporate as much detail as is necessary. For example, the compressors could be represented with all their relevant energy flows as discussed in the text.

Sankey diagram and alternative methods or increased energy conservation techniques should be considered. For the standard rate of production, the plant must be designed for minimum energy consumption. This may entail some inefficient operation when the plant is first brought into production and full design output has not been reached. Commercial considerations may make it necessary to increase production at a later date beyond the original designed rate and so provisions for increased rates without excessively increased energy consumption are essential considerations for the initial production plant designs.

Whenever a production process requires accurately controlled temperatures or humidities, this should be limited to the minimum building volume to satisfy production requirements. The remaining zones can then be less stringently controlled for occupants' comfort and therefore provide greater opportunity for energy savings.

6.6.2 Production Programming

If it is possible to arrange production with the minimum of start/stop cycles and to operate at the most energy efficient rate, this will help minimise energy consumed per unit of output. The accurate control of space temperatures and humidities might be linked to production control facilities.

Where normal operation is based on products made to order, consideration should be given to continued production after satisfying the outstanding order, the remainder being stockpiled.

6.6.3 Services for Production Requirements

The general principles of design of buildings and services for energy conservation are no different for industrial process applications from those stated previously, although process requirements are sometimes the dominating energy requirement and so deserve first consideration. For example, when there is a requirement for compressed air, heat from the aftercooler might be successfully integrated with production heat requirements. One advantage would be that, as production increases, so the compressed air consumption increases and this in turn generates more heat for the greater production rate. Hence, the heat supply remains approximately related to production rate, whereas using waste heat from compressed air plant for, say, space heating loses the benefit of this relationship and moreover there may be no space heating requirement at all during the summer months.

Industrialists must consider whether daylighting is essential for production. Rooflights can provide excellent daylighting which saves energy initially by reduced use of artificial lighting, but then can incur considerable operating costs for temperature control. Rooflights cause severe heat loss in winter in temperate and cold climates and substantial heat gain in summer under solar gain conditions. This adverse performance can be tempered by shading, protection and double skinned construction. However, complete elimination of natural lighting is not recommended. Some natural lighting is useful during failure of artificial lighting and visual communication with the outside is regarded as psychologically beneficial by providing external contact and information about weather conditions.

The building design team will require data concerning the environmental requirements of the production process so that the building and its services can be designed accordingly. Information from industrialists' past experience should be treated cautiously. For example, comments implying that a building will not require any heating often mean that inadequate insulation of production plant was previously providing copious heat for winter space heating requirements which was thrown away during the summer through louvres in the roof. Correct insulation of production plant will provide better environmental conditions for operators, at reduced costs and lower energy consumption.

6.7 CONSIDERATION OF SOME COMMERCIAL DECISIONS

Energy conservation in the commercial field requires consideration of the variety of different occupancy requirements and at what times they are needed so that the building design team can design accordingly. If similar requirements can be grouped together this may provide initial opportunities for energy savings.

6.7.1 Activity Level and Clothing

Combining people with different activity levels or differing amounts of clothing in one zone usually culminates in the environmental conditions being arranged to satisfy the most thermally sensitive, who are usually those with least clothing. A common example in cold climates is shops. Temperatures are usually selected to satisfy the staff who are clad in light clothing and often remain standing or seated. Whilst the temperature may satisfy the activity and clothing of the staff, the heavily clad shopper, perhaps carrying heavy loads, may consider the shop overheated and oppressive. Totally enclosed shopping precincts partially solve this problem by eliminating the external environment between shops.

It is generally considered that in the commercial field sedentary office workers are particularly critical of thermal conditions. This usually relates to the warmth of the surroundings but there are instances in cold or temperate climates in which sedentary office workers show a preference for a reduction in the temperature setting for their room's heating system and in compensation then wear more clothing. Such modifications may be appreciated as being beneficial because of moral obligations to reduce energy consumption, or they may be due to physiological effects such as producing a 'fresher' environment and less drying of the skin, eyes and respiratory passages due to higher relative humidities at the lower air temperatures (Section 2.2.3).

6.7.2 Occupancy Periods

If it is necessary for certain departments to operate outside the times applying to the rest of an establishment, then there could be energy saving benefits by grouping such departments together so that one zone of the building is controlled for the extra periods. This is exemplified by retail and office accommodation in one building. The retail department may operate at weekends or evenings, outside the usual office hours.

6.8 CONSIDERATION OF SOME DOMESTIC DECISIONS

The incentive for occupants to save energy in the domestic field appears to be stronger than in the commercial or industrial sectors, perhaps because the link between the consumption of energy and its cost has a more profound and immediate effect (unless there is no direct charge for the energy consumed by tenants in rented accommodation). There is often more scope for energy saving in the domestic sector because the services are usually simpler and under the control of the individual. General principles for energy saving have been widely published in many countries (e.g. [6.9, 6.39]) and the following is only a brief consideration of dynamic performance and energy conservation.

6.8.1 Average Temperature Difference

The common misconception that less energy is consumed when heating is maintained continuously has already been covered (see Section 6.4.2). Hence, the objective is to keep the difference in temperature between the controlled medium (hot water or space) and the surroundings to a minimum and for the shortest duration provided that problems such as mould growth and condensation do not become critical. When it is necessary to start up any central plant with thermal capacity (as opposed to plant with almost no thermal capacity and instantaneous maximum efficiency, e.g. electric immersion heater elements) then this should preferably be done only when there is a load for the plant. It should then be turned off when the load falls to low levels such that other plant without a load related efficiency factor might be used instead.

6.8.2 Occupancy Periods

Continuously occupied premises such as elderly persons' homes will require temperatures to be controlled throughout the occupied periods, but intermittently occupied premises can have their services operated for the minimum duration either by manual adjustments or by use of electronic optimum start and stop control systems (see Section 6.5.2).

REFERENCES

6.1 Department of the Environment: *Building (First Amendment) Regulations 1978. Part FF, Conservation of Fuel and Power in Buildings other than Dwellings.* HMSO, London, 1978.

6.2 Department of the Environment, Welsh Office: *Guidance Note: Conservation of Fuel and Power in Buildings other than Dwellings.* HMSO, London, 1979.

6.3 Peach, J.: The CIBS building energy code. *Proceedings of the Air Conditioning and Energy Conservation Conference*, University of Nottingham, 20th/21st April, 1978, Architectural Press, London, 1980.

6.4 Hitchin, E. R. and Hyde, A. J.: The estimation of heating energy use in buildings. *Symposium on the Environment Inside Buildings.* Institute of Mathematics, Southend-on-Sea, May 1979.

6.5 Billington, N. S.: Estimation of annual fuel consumption. *J. Inst. Heat. Vent. Eng.*, Vol. 34 (1966), pp. 253–256.

6.6 Chartered Institution of Building Services: *CIBS Guide B18: Owning and Operating Costs.* CIBS Publications, London, 1972.

6.7 Hyde, A. J.: Degree days for heating calculations. *Building Services and Environmental Engineer*, Vol. 2 (Feb. 1980).

6.8 Bassett, C. R. and Pritchard, M. D.: *Heating.* Longman, London, 1969.

6.9 Burberry, P.: *Building for Energy Conservation.* Architectural Press, London, 1978.

6.10 Black, A. W.: Is bigger really better? A heretical view of computer programs. *ASHRAE Journal* (Jan. 1978), pp. 48–52.

6.11 Ayres, J. M.: Predicting building energy requirements. *Energy and Buildings*, Vol. 1 (1977), pp. 11–18.

6.12 Granum, H., Markus, T. A., *et al.*: The performance and economics of single and double glazing in Britain and Scandinavia; a window chess game. *Proceedings of the CIBS Conference on Energy Conservation in the Built Environment*. Copenhagen, Vol. 6 (May 1979), p. 111.

6.13 Nicol, K.: The energy balance of an exterior window surface. *Building and Environment*, Vol. 12 (1977), pp. 215–219.

6.14 Weller, J.: *The Computation of the Energy Consumption of Buildings and its Application to the Thermal Performance of Dwellings*. Ph.D. Thesis, CNAA, Lanchester Polytechnic, Coventry, 1977.

6.15 Building energy estimation program. *Computer Aided Design*, Vol. 8(3) (July 1976), pp. 205–206.

6.16 Spielvogel, L. G.: Comparison of energy analysis computer programs. *ASHRAE Journal* (Jan. 1978), pp. 45–47.

6.17 CIBS Computer Applications Panel/Task Group: Program evaluation—building energy use estimation programs. *Building Services Engineer*, Vol. 45 (Feb. 1978), pp. 211–213.

6.18 Chartered Institution of Building Services: *Building Energy Code: Part 1, Guidance towards Energy Conserving Design of Building and Services*. CIBS Publications, London, 1977.

6.19 American Society of Heating, Refrigeration and Air-conditioning Engineers: *Energy Conservation in New Building Design—ASHRAE Standard 90–75*. ASHRAE, New York, 1977.

6.20 European Community Commission—Permanent Directory of Energy: *Information Services in the European Community*. Commission of the European Community, Brussels, 1977.

6.21 Markus, T. A. and Morris, E. N.: *Buildings, Climate and Energy*. Pitman, London, 1980.

6.22 Kasabov, G. (Ed.): *Buildings the Key to Energy Conservation*. RIBA Energy Group, London, 1979.

6.23 Sherratt, A. F. C. (Ed.): *Proceedings of the Air Conditioning and Energy Conservation Conference*, University of Nottingham, 20th/21st April 1978, Architectural Press, London, 1980.

6.24 Chartered Institution of Building Services, Institute of Energy, Royal Institution of Chartered Surveyors, Department of Energy: *Proceedings of the Experience of Energy Conservation in Buildings Conference*, University of Nottingham, 20th/21st March 1980.

6.25 St. George's School, Wallasey. *Architects' Journal* (25th June 1969).

6.26 Shaviv, E. and Shaviv, G.: Modelling the thermal performance of buildings. *Building and Environment*, Vol. 13 (1978).

6.27 Shaviv, E.: Energy conservation and solar houses. *Intern. J. Ambient Energy*, Vol. 1 (Jan. 1980), pp. 5–14.

6.28 Wilberforce, R. R.: The energy balance of glazing. *Building Services Engineer*, Vol. 43 (March 1976).

6.29 Meteorological Office, UK: *Tables of Surface Wind Speed and Direction over the United Kingdom*. HMSO, London, 1968.

6.30 Chartered Institution of Building Services: *Building Energy Code: Part III, Guidance Towards Energy Conserving Operation of Buildings and Services*. CIBS Publications, London, 1979.

6.31 BRITISH GAS CORPORATION: *Energy Demand and System Sizing*. British Gas Corporation, London, 1980.
6.32 SPRAGUE, C. H. and MCNALL, P. E.: The effects of fluctuating temperature and relative humidity on the thermal sensation (thermal comfort) of sedentary subjects. *ASHRAE Trans.*, Vol. 76(1) (1970).
6.33 FANGER, P. O.: *Thermal Comfort: Analysis and Applications in Environmental Engineering*. McGraw-Hill, New York, 1970.
6.34 HITCHIN, E. R.: The sizing of heating systems for well insulated homes. *Applied Energy*, Vol. 5 (1979), pp. 277–295.
6.35 BLOOMFIELD, D. P. and FISK, D. J.: *The Optimisation of Intermittent Heating*. Building Research Establishment Current Paper, CP 14/77, 1977.
6.36 WHITTLE, G. E. and WARREN, P. R.: *The Efficiency of Domestic Hot Water Production out of the Heating Season*. Building Research Establishment Current Paper, CP 44/78, 1978.
6.37 REAY, D. A.: *Industrial Energy Conservation*. Pergamon Press, Oxford, 1977.
6.38 PAYNE, G. A.: *The Energy Managers' Handbook*. IPC Press, Guildford, 1977.
6.39 BUILDING RESEARCH ESTABLISHMENT WORKING PARTY: *Energy Conservation: A Study of Energy Consumption in Buildings and Possible Means of Saving Energy in Housing*. Building Research Establishment Current Paper, CP 56/75, 1975.
6.40 OWENS, P. G. T.: Figuring out energy needs, *CIBS J.*, Vol. 2(7) (July 1980), pp. 43–45.
6.41 BURGESS, K. S.: *Computer Programs for Energy in Buildings*, Design Office Consortium Evaluation Report no. 5, 1979.
6.42 WATERS, J. R.: The experimental verification of a computerised thermal model for buildings, *Building Services Environmental Engineering Research and Technology*, Vol. 1(2) (1980).
6.43 REAY, D. A.: *Heat Recovery Systems: A Directory of Equipment and Techniques*. E. & F. N. Spon, London, 1980.

Chapter 7

Design Team Approach

7.1 INTRODUCTION

There is no standardised approach to a project by a design team. If the same brief was given to two separate design teams, two distinctly different solutions are likely to be produced. The basic requirements of the brief, such as the necessary floor area, would be met but the solution to many of the detailed problems would be different. There are many reasons for this; some are now given to introduce the topic of the design team approach.

7.1.1 Who Should be Appointed?

The variety of expertise which may be required for a project will depend on the specific project requirements, although all projects will require expertise for the fundamental design functions of architecture, engineering and economics. On large projects, this results in the appointment of architects, civil engineers, structural engineers, services engineers, building economists, quantity surveyors, energy consultants and a design team manager. In addition it may be necessary to appoint specialists to take responsibility for specific subjects which are beyond the depth of understanding of those already appointed to the design team. As many of these design team members will be specifying particular materials or systems, it will be necessary to obtain the observations and approval from the manufacturer of the materials or systems for many applications and this process introduces further expertise into the project. Small projects may require the same variety of expertise as the complexity of problems is not necessarily related to the size of projects, but the variety of expertise is sometimes restricted by economic limitations. Previous experience usually speeds progress through early design stages and so experienced designers will be of greater value to a design team.

The selection of personnel for the design team involves all the usual management considerations such as education, experience, compatibility with other members of the design team, flexibility of thinking, vested interests and enthusiasm for a particular project. Careful consideration must be given to the appointment of any designers who might have their design flexibility inhibited by, for example, preconceived ideas from previous experience or being obliged to specify only their employer's products, particularly if that employer does not market a range sufficiently wide enough to cover the variety of possible options.

A typical case is the expertise which many designers have developed in the field of warm air heating installations, whereas a radiant heating installation may be a more energy conserving solution to the brief. Another example might be the need to consider reduction of solar heat gain; would a glazing manufacturer recommend external louvres or a louvre manufacturer recommend structural projections? Salesmen are trained to deal with awkward questions and to avoid highlighting the disadvantages of their products. Good salesmen appreciate that the short term financial reward for selling their company's products in an incorrect application may jeopardise long term business and so they may recommend that another solution to the problem is found as their range of products is not suited.

A successful appointment to the design team requires a person with an unbiased approach who has the flexibility and expertise to make a useful contribution to the design as it progresses through all the various stages, although at specific stages it may also be necessary to appoint specialists to deal with detailed problems. The input from specialists must be co-ordinated by a member of the design team who must be able to discuss details with the specialists and ensure that the final solution is the most satisfactory, after all the specialist aspects have been considered.

For the design of buildings in an era when energy conservation measures are becoming more important, it is essential that the designers are aware of the latest products, techniques and ideas. This must be combined with experience so that a balanced approach is obtained. It is therefore reasonable to expect designers to have links with educational establishments, research organisations, information centres and professional institutions, and to be well informed about the current thinking on topical subjects. This requires an investment of time, in addition to the time allocated to projects, for reading publications, and for attending seminars, conferences, mid-career courses and for visiting other projects.

7.1.2 Remuneration

The reward for any input helping the successful progression of the project must be carefully arranged to prevent any influence which may be detrimental to the overall design. The following discussion is aimed at drawing attention to some of the incentives which are inherent in some of the fee structures.

Any design process will consume time, effort and money. The 'free' design service which can sometimes be obtained from manufacturers, installers and fuel or energy supply industries must be used with caution because of the obvious vested interests. These interests may include selection of materials and plant, speed of installation of various systems and the economy (or lack of economy) of operation. However, this is not to say that such expertise should not be consulted. In fact, invited comments are sometimes essential to ensure successful design, one example being the potentially destructive thermal stresses in solar control glazing which might be caused by partial shading of large panes. Direct approval of an application by the manufacturer is then valuable.

When it is intended that a 'free' design service is also combined with competition based on total costs of design and installation, then the emphasis is placed on designing as fast as possible the cheapest installation which will just achieve the explicit requirements of the brief. Any extra design time for a different solution to the brief must produce a capital saving in excess of the cost of the extra design time. This technique of 'design and construct' is generally accepted as one which produces relatively poor designs where energy conservation is desirable, and not necessarily at the lowest capital cost because contractors must include the cost of preparing designs for their unsuccessful competitive quotations which may typically number seven for every one successful quotation. During such projects any modification to the contract might be used as an opportunity to maximise profits.

The appointment of unbiased designers who are rewarded by fee scales based on a percentage of capital costs can also influence designs. If the fee scale is directly related to the capital cost of the project there is a positive disincentive to the designer to put extra effort into investigating and specifying another system with a comparable performance but a lower capital cost, simply because the extra effort is penalised by reduced remuneration. This is exemplified by the occasions when a modest investment in solar control produces a greater saving in air conditioning and refrigeration plant costs. Furthermore, there is no direct incentive for a designer to put extra effort into a design which operates at a lower energy consumption. There is only the indirect benefit of improved goodwill and prospects of future work when the designer's name is associated with complimentary comments about the design. However, such compliments are limited because each building design is unique and so comparisons cannot easily be made, and energy control is not usually the responsibility of any single member of the design team—it may involve the architect, interior designer, environmental engineer and services engineer. Hence, it is difficult for even the discerning occupier to appreciate that savings in operating costs are actually being achieved, but when such savings are appreciated, compliments are usually credited to the design team as a whole.

Another form of remuneration is to pay designers according to the time involved. 'Payment by the hour' can lead to problems of lack of financial control because there is no incentive to work diligently, except that a poor reputation may be earnt by the less industrious designers.

In other situations, the traditional solution to problems of remuneration is to offer payment by results, but this does not work easily in the context of energy conservation. The 'results' could be interpreted as minimal energy consumption, achieving energy targets or overall economic considerations; but whichever approach is chosen there is still the problem of deciding what benefit the designer has been (and therefore what remuneration should be offered), because there is often nothing available for comparison purposes, except with the conversion or improvement of existing buildings.

Payment by agreeing a single fee for a project tends to leave only one incentive; that is to complete the work quickly so that another project can be started. This does not necessarily cause any problems but this method does place the emphasis on the designer's previous experience to predict the amount of work which will have to be done in order to come to an agreement about fees. It would be natural to expect an inexperienced designer to err on the expensive side to cover himself against unforeseen circumstances.

7.1.3 Briefing the Design Team

The brief is of particular importance to the design team and so its preparation deserves the maximum possible input. Preferably, it should not stipulate the format of the design but instead should state lucidly the basis for the design. It is the foundation from which all design discussions will commence and so it will intrinsically impose a significant influence on the outcome of the design process.

In relation to energy conservation, an attempt should be made to include the following general information in the brief:

(a) The expected duration for which the controlled internal environment may be required. This may be influenced by changes to production or occupancy which could occur every year or so.

(b) The duration for which the building envelope will be required. The building envelope may remain through many changes to the internal environment.

(c) Having determined from (a) and (b) information on the design life, decisions must be made, usually based on economics, using data which must be supplied to the design team (see Section 7.2).

More definitive information and technical information on specific requirements should be written in such a way as to produce the least constraint on design. This will provide the design team with an opportunity to consider solutions to the brief without having limitations imposed or receiving preconceived ideas. Technical information should be carefully vetted before its incorporation into the brief to ensure that it is correct in every respect. Even information gained from experience should be treated with caution. The experience may have been obtained from circumstances different to the current project and so information derived from previous experience should be supported by background information to ensure it is used correctly.

Some technical information, such as the temperature and humidities required for an office environment, is intuitively apparent to the design team, whereas for example the detailed environmental requirements for production or storage of materials may require specialised knowledge. Such information should be provided when it is readily available; alternatively references should be made to appropriate sources when the information is not immediately available. In general applications, the information which the design team will require commences with the degree of control necessary and the duration for which that control will be required. Further requirements will relate to each specific design problem.

The preparation of the brief demands expertise and experience. In fact, professional advice can be invaluable during its preparation in order to ensure that the correct information is provided in the right manner. After all, the brief is the foundation for a project which may be extant fifty years after the design has been completed.

7.1.4 Contractual Arrangements

Previous discussion has already mentioned the competitive 'design and construct' quotations which inherently incur incentives to minimise design time and construction costs, generally at the expense of energy conservation. With 'design and construct' contractual arrangements there might be the incentive to minimise the cost of the members of the design team, by employing less expensive designers who may not have the required education, training and experience to produce the best solution to the brief. Alternatively, more expensive designers may be able to produce better designs at lower construction cost due to their greater expertise. In either case there is little control over the quality of design because the appointment of the members of the design team rests with the competing 'design and construct' contractors.

There are many different forms of contract and it is not proposed to examine each one. However, it is suggested that the time restrictions due to the element of competition, and the disincentives caused by fees which are directly related to capital costs, both reduce the possibility of the best designs developing from the brief. Hence, to overcome these difficulties it would be necessary to divorce competition from the design and to reward the design team fairly for the work which is necessary.

7.2 ECONOMIC ANALYSES FOR INVESTMENT DECISIONS

7.2.1 Introduction

The incentive to conserve energy comes directly from the saving in operating costs and sometimes indirectly through reduced size of plant, reduced reliance upon mechanical services, reduced availability of energy in the future, reduced maintenance and improved morale. Occasionally, there are other reasons for energy conservation such as a need to limit demands so that limited waste energy from a production process can completely satisfy a known demand, or the extra status which may be obtained from owning a system which is novel, innovative and morally justified. Any of these reasons may be taken into account in the analysis of whether to invest in an energy conserving measure.

In order to assess the true economic value of any investment in conservation measures, the value of the benefits during the life of the building and the plant must exceed the value of the additional investment, otherwise there will be no incentive to invest. Regretfully, such incentives are of no value when capital is provided by one organisation or department, and revenue is provided by another. Common examples are government departments and rented accommodation. This is one reason why some form of legislation to conserve energy is desirable.

There are many methods of economic analysis of investment decisions. Economists, accountants, company directors, designers and lay people all appear to have their preferences. Whilst some of the techniques may be adequate for certain straightforward investment decisions, some have flaws which make them inappropriate for making decisions about energy conservation options. This section presents the merits and demerits of a few of the techniques; further information can be obtained elsewhere, e.g. [7.1, 7.2, 7.3].

All the techniques require the prediction of the value of the gross energy savings; hence the need for the prediction of energy consumption as discussed in Chapter 6. The value of any other merits may be incorporated if they are considered appropriate.

The value of the demerits must also be incorporated. These may include:

(a) The lost interest on the capital previously invested elsewhere, or the additional interest on the loan for the extra investment.
(b) Any additional energy consumption, e.g. the running of heat recovery fans.
(c) Additional maintenance for the extra equipment.

7.2.2 Simple Payback Period

One of the simplest methods of investment decision making is to calculate the period for which the savings will have to accumulate in order to equal the extra investment. If this period is longer than the life of the investment, obviously there is no incentive to invest. As this technique is usually applied to short life investments, inflation or deflation is usually ignored.

Unfortunately, the technique may not give indications of the best investment when there is a choice available and hence the wrong conclusions may be drawn. For example, consider an investment decision regarding two mutually exclusive options for reduced energy consumption by production plant with a guaranteed life of four years. Option A costs £1000 and saves £1000 per year overall when allowing for the lost interest on the capital or the extra interest on a loan (payback period 1 year). Option B costs £2000 and saves £1500 overall per year (payback period 1·33 years). Option A may be selected because of the shorter payback period but further analysis reveals:

Option A: £4000 saved over 4 years less £1000 capital = £3000
Option B: £6000 saved over 4 years less £2000 capital = £4000

Although option B has a longer payback period it produces a greater overall saving at the end of the investment period when the capital is repaid out of the overall savings. Hence, it is clear that simple payback periods do not provide a full analysis for decision making. Furthermore, no allowance has been made for inflation, deflation, changing energy values or the timing of the returns.

7.2.3 Simple Return on Investment

The average net annual savings (return) are calculated and related to the extra capital investment which is then expressed as a percentage. If this return on investment exceeds the lost interest on the capital or the interest paid on the extra loan then the investment is potentially attractive.

This technique is usually applied to long term investments (e.g. 20 years) and therefore the changing value of money and energy should each be taken into account, as well as the terminal value of the investment or the total value of the savings.

However, the technique does not facilitate this more comprehensive form of analysis and so its popularity is restricted.

7.2.4 Accumulated Cost Comparisons

When it is necessary to consider a variety of systems with different capital and operating costs, one technique is to consider the accumulation of capital and predicted operating costs as time progresses. The information is usually presented graphically as illustrated in Fig. 7.1.

Figure 7.1 shows four options. Option A is the cheapest to install but the

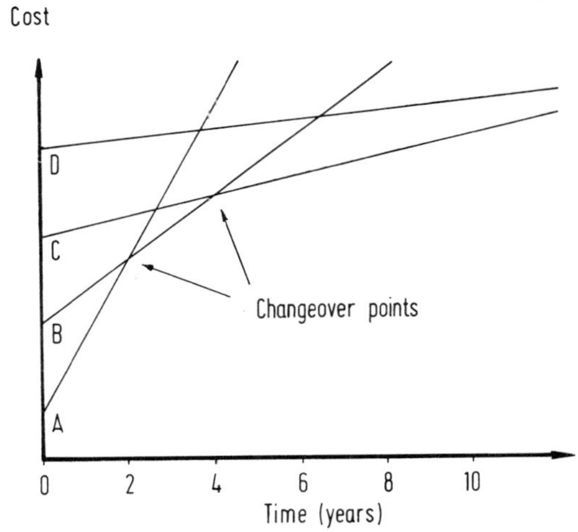

Fig. 7.1. Accumulated cost comparisons of four investment options, A, B, C and D.

most expensive to run. It accumulates the minimum total cost if it is only required to operate for a period of less than two years. Between two and four years option B has the cheapest accumulated cost and so there would be a change of selection at the changeover points indicated in Fig. 7.1. Although option D has the lowest running costs, it is penalised by the highest capital cost and does not become attractive before the predicted end of the life of the investment (10 years). Hence, option C is the better long term investment.

Allowances can be made for major maintenance items which may be required at various intervals with each option, and changing energy values can also be incorporated. However, no allowance has been made for the reduced value of future cash flows.

7.2.5 Discounted Cash Flow Methods

None of the previous techniques have inherently made allowances for changing values of money or energy and so the true value of an investment may not have been presented. The following example is a simple case which is exaggerated to illustrate the point. Consider taking money out of a previous financial investment earning interest at 10 % per annum on the capital invested. Instead, it is invested in thermal insulation which also earns 10 % per annum due to reduced energy consumption. Assume that the value of energy now rises during the investment at 20 % a year and so the return on the initial investment as time progresses can be compared thus:

Year	Financial investment return (%)	Thermal insulation investment return (%)
1	10	10·0
2	10	12·0
3	10	14·4
4	10	17·3
5	10	20·7

Hence, after five years, the insulation is producing over 20 % annual rate of return on the initial investment whereas the financial investment would still be returning 10 %. If the value of money had also fallen during the period at 20 % per annum (inflation) then the value of the insulation would remain the same in terms of energy saving whereas the actual value of the 10 % interest on the financial investment would be substantially diminished.

It is clear that if the energy saving benefits of the insulation are only considered in monetary terms and money changes value as well, there are two events occurring simultaneously; the changing value of energy as easily obtainable world resources diminish, and the changing value of money (inflation or deflation). This problem is sometimes expressed by economists who differentiate between a 'real' value and an 'apparent' value when inflation is ignored.

To clarify the problem it is necessary to consider the present value of money which will be obtained in the future. Consider the present value of a promise of £1 in one year's time. To obtain £1 in one year's time a sum of money would have to be invested now at the predicted interest rate during the forthcoming year so that the total of the original sum and the interest amounts to £1 after one year. Clearly, the original sum must be less than £1 and so a promise of £1 in one year's time is worth less than £1 today. The larger the interest rate, the less the present value of future income and so it is important to weight the value of future income during periods of high interest rates.

Any investment in energy conservation measures, whether by improved design of the building, additional equipment or additional insulation, produces a return in the form of reduced energy consumption. As time progresses, so the value of money changes with inflation or deflation, but the amount of energy saved will usually remain constant (unless equipment performance decreases with age) and so the actual financial return will vary with changing energy values. Hence, high rates of general inflation encourage investment in energy conservation measures, and increasing value of energy produces greater financial return from such measures.

Some methods of economic analysis discount all future cash flows back to a base date which is used to make value judgements. Often the value of money at the present day is chosen and this is called 'present values'. When the cash flow varies with time (e.g. a major maintenance item after 10 years) then these have to be separately discounted back to the base date. Investment decisions can be assisted using either of the following techniques:

(*a*) *Net present values*. Capital investment, annuities and any other cash flows are discounted at the selected rate and the sum of all the present values gives the net present value. Any outlay is deducted and any return is added to give the net present value. If the answer is positive the investment is worthwhile, and if zero or negative the investment is not worthwhile. Several options can be compared to find the best.

(*b*) *Discounted yield or rate of internal return*. This technique uses a trial and error method of calculating the interest rate which will have to occur to make the net present value of the investment amount to zero. This interest rate is called the discounted yield, and to arrive at a decision it is compared with the interest rate on the capital previously invested or the interest rate which must be paid on the loan for the project. If the discounted yield is greater than either of these interest rates, the investment is worthwhile.

The discounted yield technique is usually preferred because it shows the rate of return on the investment per annum. When capital is limited (which it usually is) then maximum profit can be obtained by investing in the high rate options. When energy saving equipment is becoming old and more money is needed to be spent on maintenance (i.e. negative cash flows), this technique can be used to decide whether the present equipment remains in use or is scrapped and replaced. However, this technique is rather more time consuming to calculate by hand although programmable calculators can assist and save time.

Although the more sophisticated methods of economic analysis produce safer and more logical answers to the simple question, 'Is it worth investing in this energy conservation option?', there is still a problem. Predictions must be made of how much the value of energy will rise year by year during the economic life of a project. Note that the economic life of buildings is often between 30 years and 50 years! Hence, decisions on conservation measures inherent in the design of the building or within the construction involve these long term predictions whereas decisions regarding conservation measures inherent in the building's services are often based on the assumption that their economic life will probably not exceed twenty years.

7.3 DESIGN TEAM CONFIDENCE

Even though an economic analysis has shown the favourability of a particular option, the confidence of the design team may be undermined by influences which cannot easily be valued. The following illustrates the type of problem which may occur.

7.3.1 Technical Innovations

Manufacturers respond to changing market demands. A sudden demand in a particular field previously little researched can produce two types of reaction initially. The first is the production of very simple equipment which is perhaps rushed onto the market and is supported by advertising which emphasises the simple nature of the device. Whilst simplicity is not condemned on principle, sometimes the degree of simplicity suggests that the device may not be able to achieve the claimed performance and there is often insufficient supporting research.

The second reaction is often called the 'second generation' and relies upon existing technology or equipment already available on the market. A small amount of research will have been carried out to add additional components or modify some of the existing components so that the hybrid plant can perform a new duty. This causes concern due to the original items and new production items coming together to perform a new range of duties. It is always difficult to predict the reliability and economic life of such arrangements. Sometimes the technical complexity of the original equipment may have to be made even more complex to enable the plant to undertake new duties. This can lead to problems due to lack of understanding by the user or maintenance staff.

After the initial reactions, the equipment specifically designed for the new tasks becomes available and this naturally undergoes development as more experience is gained from its operation. A designer must be aware of the stage of development an option has reached. Before there is any investment by manufacturing industries in the development of new technology, there must either be an existing demand for a particular product or it must be confidently predicted that there will be a demand for the product when it becomes available. Consider the following simplified presentation of the economic influences:

(a) The production quantities will be geared to the predicted demand.
(b) The demand will vary with the price of a product.
(c) The price will vary with the production quantities.

The demand for energy conservation measures is triggered by the value of energy. As the value of energy increases, there is an increase in the demand for energy conservation measures. As production is increased to meet the demand, the price of the energy saving measure decreases and this makes the investment in such measures more attractive and so the demand increases further.

7.3.2 Market Acceptance

When technical innovation has progressed sufficiently, the market support begins to grow. However, on occasions, the market support for some items of equipment or design features does not grow and the items are withdrawn from the market. Maintenance and repair at a later date may then cause problems.

7.3.3 Support During Design

Obtaining data during early exploratory stages of design and obtaining detailed technical data during the design are sometimes difficult. This may be due to a wide variety of reasons, such as the policy of some companies not to provide support, except for sales literature, until they have received an order. However, as the quality of the design process improves, the input of technical data must be of consistent quality and so it must improve in sympathy.

7.3.4 Quality of Design Staff

Whilst the 'shop window' of any design organisation may be technically competent, well qualified and with useful experience, this must be backed by design staff who are able to carry out all the details of the design to the level of quality which is inherently essential. Whilst it may be prudent not to accept commissions which are beyond the level of competence of the design office, the need for work and other pressures may encourage the acceptance of commissions which are too complex for the design staff to undertake competently. This leads to consideration of only those options which can be designed successfully by the design staff; the remainder may be unjustifiably abandoned or rejected.

If an attempt is made to operate outside confidence limits, the design staff may add 'safety margins' to their design to cover for their inexperience and lack of knowledge. This can be particularly destructive to energy conservation measures because the capital cost is unnecessarily increased. Furthermore, over-sized plant will spend more of its time operating well below its maximum duty and this can lead to problems of inefficiency, more frequent breakdowns and increased maintenance.

The problem of obtaining adequate design staff is compounded when the industry's recruitment is low or the education of the newcomers is poor. In addition, any change in demand which creates a need for a different approach to a design problem or a new level of thinking to obtain adequate solutions can cause difficulties for experienced designers who may have become inflexible and entrenched in their ways. This creates a need for 'mid-career training' so that design staff can respond with a flexible and innovative approach to the new demands placed upon them.

The commercial aspects of business management are arranged to provide some degree of security for the occasions when a design fails. When the design is provided within the contract with a company of limited liability, the extent of that liability will depend on the details of the contract conditions or on the company's limited liability. Care must be exercised when entering design contracts which are for values in excess of the company's limited liability and sometimes a bond is held by a third party to provide some security against inadequate performance. In fact, some organisations and government bodies are loath to enter such contracts and so any design contract must be backed by unlimited liability. Design and consultancy practices which offer unlimited liabilities usually have an insurance policy which provides professional indemnity. This produces the irony of unlimited liability for the design although the contractor responsible for the installation is limited in liability. However, professional indemnity does not resolve problems caused by incompetent design when the design aims to provide an energy conservative approach, unless the design has clearly failed to achieve a particular performance.

7.3.5 Speed of Design and Construction

Even though competent experienced design staff may be available for a design, some designs inherently take longer to complete than others, especially when the design is somewhat innovative.

There may be a need to collect information from research bodies, educational institutions (universities, polytechnics, etc.), information centres or from visits to previous projects, where it may also be necessary to measure performance and collect data over a period of time. If there is no previous knowledge, it may be necessary to initiate specific research projects to investigate the problem and increase the depth or breadth of knowledge. There are a wide variety of research establishments and educational establishments where research projects can be undertaken. However, the design time must be lengthened to allow sufficient time for the research project itself to be designed, programmed, set up, operated, concluded and reported, whilst members of the design team will be involved in directing and monitoring the research work.

The amount of time required for a design may also be increased because of a need for more calculations or for calculations with increased detail. This is usually accompanied by a demand for more information from manufacturers of equipment or suppliers of systems, which may require more time, and the actual process of drawing up the design and preparing the specification may also be more involved and time consuming.

Any constraint on the time available to the design team, such as a building which must be available in an unreasonably short time, will of necessity curtail the amount of work which the design team can complete and will probably influence the design, forcing it in the direction of previous experience and to schemes which are quick to design and construct. In order to maintain confidence at being able to complete the design and allow sufficient time for the contractors to construct the building, the design team may choose to ignore more complex schemes and sacrifice potential energy savings, even though the economic analysis concluded that such schemes were worthwhile.

7.4 THE USER'S UNDERSTANDING AND ACCEPTANCE OF SYSTEMS

Energy conservation systems may be wasted if the user of the system cannot respond knowledgeably to the demands which the system may explicitly or implicitly place upon him. For example, one incidence of poor temperature control was caused by the office manager who used the wall mounted room air thermostat as a hanger for his hat and coat. The implicit system demands were not met!

The following illustrates some of the consequences that can arise from decisions made by design teams.

7.4.1 Decision Making by the User

If it is possible to design a system to be fully automatic, so that any complex logic is resolved inside the enigmatic 'black box', then this is acceptable providing those responsible for maintaining the system and the black box are able to resolve any system malfunctions caused by errors in logic. However, sometimes the logic becomes so complex that this causes concern to the designer or the decision making process cannot be resolved satisfactorily by logic. This is exemplified by the psychological factors influencing whether motorised solar shading devices should be withdrawn to provide a view for the occupants. It may be considered easier and cheaper to design a complex system for manual control, provided the operator understands the decision making logic and is prepared to accept the responsibility for continuous attention and operation.

Some decisions require a higher level of understanding than is common amongst system operators. Consider the situation of someone faced with an apparently simple choice of heating water in a home in summer by:

(a) An electric kettle.
(b) An electric immersion heater in a hot water storage tank.
(c) An electric immersion heater in a hot water storage tank using off-peak tariff electricity.
(d) The central heating installation consuming a fossil fuel.

In order to advise which method is cheapest to run, a series of questions must be answered, the more important ones being:

(a) How much hot water is required?
(b) What is the relative purchase price of each form of energy?
(c) What is the actual efficiency of each system and therefore the actual cost of obtaining the heated water at the point of use?

The answer to the last question varies with the quantity of hot water required, which also affects the efficiency of each system. Hence, there is no generalised conclusion to the questions and therefore no generalised recommendations.

Such problems become more complex as the complexity of the system increases. For example, the operators of an old group heating system were faced with similar problems but of a much greater cost implication. The group heating system served many buildings on a site. During modernisation of the buildings, supplementary heating facilities had been installed in each building to augment the group heating system. The system operators now face the task of assessing which system will operate at least cost under a wide variety of different conditions.

The number of options facing system operators increases with system complexity. The selection of the most efficient method of operation for any set of conditions requires data which are usually not measured. An estimate of the system efficiency could be obtained by computer simulation or even by manual calculations, although both are usually beyond the ability of the system operators who are having to make the decision. The installation of instrumentation to measure a system's efficiency is usually not easy and so these decisions are often left to intuition.

However, developments in the application of computers to the control of building services have provided the potential for continuous monitoring of the performance of systems. The information must be analysed by the system operators who can then make decisions on the best method of operating the system. A more sophisticated application of computers involves the direct control of the mechanical services by the computer, although this requires knowledge of the mathematical interrelations (algorithms) occurring within the systems. This is often a two stage process as the development of an algorithm for any system stems from the data collected during the monitoring period.

7.4.2 The Domestic Situation

Although the general consensus of opinion is that occupants learn to use the facilities provided for them, experience suggests that this may not necessarily be true. In one situation, complaints about heating failures were traced to the timeswitch having turned off the heating. Simple adjustment of the timeswitch provided the heating when the occupant required it. In another example, a complaint about inadequate heating was caused by the room air thermostat being set too low. In both situations and in other similar cases, the occupants have stated that they do not touch the controls, suggesting a lack of understanding and some apprehension. Manufacturers' operating instructions were provided in each case.

Misconception also causes energy wastage. Heating timeswitches have been found set to continuous operation because it was considered less energy consuming than intermittent operation. Thermostatic radiator valves are sometimes turned from one extreme to the other and so behave as ordinary on/off radiator valves, instead of controlling the flow of hot water to the radiator according to the air temperature at the sensing head.

Whilst such controls may theoretically be worthwhile, it is essential to ensure that the controls which are made available for the occupants to use are actually used sensibly.

7.4.3 The Office and Industrial Situation

Small commercial firms fall into the domestic category because they do not usually have a janitor. Large firms often have a janitor, who may be supervised by an energy manager having responsibility for all aspects of energy conservation.

The energy manager's task will include ensuring that those responsible for operating energy consuming plant are made aware of the consumption characteristics and the action which the operator must take to ensure that energy wastage is minimised. In order to accomplish this, the energy manager will have to obtain a considerable amount of data and then present only the relevant data succinctly to the operator. Both energy managers and operators may require special training courses as energy management is difficult to learn simply from experience or intuition, except in simple cases. For complex situations, energy managers should not hesitate to obtain expert assistance from energy consultants.

Many aspects of successful energy management stem from the understanding of topics covered in the previous chapters. Some firms have been very successful with energy management and are seen to operate their plant most frugally. In such situations, a design team would not hesitate to provide them with complex energy conservation systems as it would be reasonable to expect the new systems to be operated efficiently. Regretfully, experience has shown that some firms have little inherent respect for energy conservation. They do not have an energy manager and the plant operators are not educated about energy conservation. Sometimes this is caused by the firm operating in a particularly lucrative market and energy costs are a small proportion of total operating costs. Consequently, they do not feel the need for energy conservation. Nevertheless, similar investment decisions apply to energy conservation measures as to any industrial investment project, and so profit could be increased further. However, if a design team provided such a firm with energy conservation systems, the firm must be made aware of the benefits and accept the responsibility for operating the system, otherwise it may fall into neglect and disrepair.

7.4.4 Illicit Tampering

When occupants do not have to pay for the energy consumed by a system, tampering with the system controls can become a problem. To prevent this, some manufacturers have introduced controls which are described as 'tamper-proof', 'tamper-resistant' or 'lockable', but occupants sometimes find ways around these restrictions. For example, in a home for elderly persons, the tenants discovered that the thermostatic radiator valves could be adjusted to a temperature above the restricted upper limit by partially unscrewing the valve head.

In Section 5.3.8 it was explained that the location of sensors is important for effective energy conservation. This usually results in sensing devices located in the occupied zone. However, the adjustment for temperature or humidity must be located remotely, with access restricted to responsible personnel, otherwise the energy savings may not be obtained due to unauthorised adjustment.

Systems provided with automatic controls are sometimes operated manually because of a desire to have personal control. Attempts to overcome this problem have included dummy controls but it is doubtful whether the absence of system response to any tinkering would remain unnoticed for long.

7.4.5 Complexity of Operation

It is clear from previous discussion that the complexity of operation is sometimes a compromise between the sophisticated complexity for efficient energy conservation under all situations and the low degree of complexity which can be easily accepted by the system operators.

On some occasions, sophisticated energy conservation measures may be beyond the scope of occupants. The system may then fall into misuse and so the potential energy savings will not be obtained. The design team must respond to this problem and design a system which will operate effectively with the understanding, acceptance and support of the occupants.

Whilst there may be the appropriate expertise within the design team and also enthusiastic system operators so that complex operation is viable, the design team must still give consideration to maintenance of the system at some stage in the future when the original enthusiasm and understanding has long since passed. The system operators may change, the design engineers may no longer be with the design firm, the installing engineers may no longer be with the contractors and so another company may be asked to deal with a malfunction in the system. Although the installation drawings may indicate the various services and their routes, and the operating instructions provide for starting, normal running and stopping the installation, it is rare to find any information which gives the reason for the presence of an item or its operational logic. This is probably because there has previously been little need for such documentation as the design of the systems has fallen into a category which is quickly recognised and understood by any experienced engineer visiting a conventional services installation. As the complexity increases and moreover as the functions become more varied and obscure, so the need for comprehensive documentation increases. This must be presented in such a manner that a newcomer will be able to gain both a general understanding of the variety of system functions and a more detailed understanding of the logic of operation of principal components.

7.4.6 Moral Aspects

It can be of value if building occupants and operators of systems are convinced that any energy conservation measure is desirable and correctly designed. In one instance, lack of conviction over a design led to problems with the heating installation in one of two similar buildings in which improvements were made to the heating installations. The first building had a gas fired heating installation. The second building was converted to an automatic coal fired heating installation. The coal fired installation repeatedly gave trouble which was eventually traced to the janitor who was sabotaging the installation. The janitor considered that this building should also have had a gas fired installation which he regarded as clean and modern and therefore the correct course of action.

On several occasions, companies have noticed an improvement in morale amongst staff when, upon the advice of an energy consultant, additional measures for energy conservation have been implemented. This improvement in morale might be due to the staff considering that the company are making the right decisions and are taking the correct course of action.

7.4.7 Balanced Quality of Environmental Control

It is important to ensure that the quality of environmental control is fairly consistent throughout all factors which influence an individual's assessment of the internal environment. Usually this is fairly easy to achieve when new buildings are being designed but greater care is needed when existing buildings are being upgraded.

For example, consider an old established factory with large office accommodation created out of the production space by a simple industrial partitioning system. The management appreciate the office staff's genuine concern over the hot, unpleasant and fatiguing conditions in the offices and so they intend to deal with these aspects, hopefully to improve morale.

An air conditioning system is designed and this incorporates heat exchangers between the exhaust and fresh air ducts to minimise running costs, filters to reduce the quantity of airborne contaminants, negative ion generators to redress the imbalance of the air ions and humidification to mitigate the drying of eyes, noses and throats during the winter. The offices will be slightly pressurised by the fresh air system so that the ingress of odours from the factory production space is impeded. The management approve the design, the installation is completed and commissioned.

The subjective response to such improvements is difficult to predict. Whilst there may be fewer justified complaints about inadequate temperatures and ventilation other influences may also be of concern. Examples are:

(a) Whilst the previous atmospheric conditions might have caused lethargy, the level of fatigue might not have reduced with the improvements because the intrusion of noise from adjacent production might still be hampering the working environment.
(b) The lack of interest and colour in the surroundings and an absence of an external view might be causing depression which in turn reduces office efficiency.
(c) Poor office management may cause antagonism and this will have a far stronger influence on the office efficiency than the benefits of the air conditioning system.

The outcome of the above may be unrelated to the actual cause. In fact, there may be complaints about the new air conditioning system and these may be directed at draughts, temperatures or smells with little or no justification. The basic dissatisfaction may be due to other factors such as noise, colour, human relationships or problems of welfare such as the arrangements for and quality of the refreshments.

Whilst it is often cheaper, in such instances, to complete all the work on the air handling system in one contract, experience has shown that the overall benefits are greater if improvements occur gradually to all the environmental factors which may be causing dissatisfaction with the environment. Furthermore, there is an advantage in tackling one problem at a time; the occupants appreciate the management's continuing consideration of the quality of the working environment and this appears to create improved morale. Hence, it is considered that gradual, continuous improvement is better than substantial improvement which only deals with one field of environmental problems. The working environment requires continuous consideration so that new research findings and feedback from personnel can be incorporated into any modifications, to demonstrate an acceptance that the quality of the working environment matters.

For the example quoted, the first stage improvement might be the installation of ductwork with fresh air, exhaust facilities and cooling only. Provisions would then be made for the other facilities at a later date after improvements have been made to the acoustical environment, the colour scheme, refreshment facilities, etc., in an appropriate order or priority. Staff comments are useful when ranking the priorities.

The need to maintain a balance in the quality of the environmental control can occasionally oppose an incentive to install energy conserving systems. This is exemplified by the preference for openable windows in clean temperate climates so that ventilation is under the control of the occupants. The open window also provides pleasures such as the sound of the birds and the whistle of the wind, in addition to the effects on air quality discussed in Section 2.4. If heat exchange between fresh air and exhaust is to be incorporated for energy conservation, then it will be necessary to forego some direct environmental benefits in order to obtain the energy savings.

7.5 DESIGN TEAM OPERATION

7.5.1 Introduction

Whom the client chooses for the position of project manager will have a significant influence on the design which evolves. Traditionally, it was usual for the architect to control projects during design and construction. But gradually there has been a change to the architect confining his input so that he can concentrate more on the design, where his skills of integration of the requirements of each expert in the design team have had to develop in response to the need for improved design of buildings.

The client now has more choice of how a project shall operate. The method which the client chooses will affect the manner in which the design proceeds and the emphasis placed on each aspect of the brief. Traditionally, most projects were initiated by the client approaching an architect for a design. The architect then sent the almost completed design to engineers but the engineers saw the project at too late a stage for their expertise and experience to be of benefit, except in the details of the service design (which might already have been constrained by architectural influences anyway). This problem was exemplified by the recurring failures through condensation within flat roof constructions in temperate climates. The technical reasons for interstitial condensation during winter were known and there were a variety of solutions to the problem, yet the high number of failures continued long after the understanding was developed. Contributing to the causes of this particular design failure were:

(a) Nobody with an understanding of psychrometrics, moisture movement through materials, transient thermal performance of materials in construction and the architectural needs was involved in the project early enough, or in some instances was never consulted at all, especially for small projects.
(b) Because flat roof construction was a cheap form of construction, it was not expected to have a good life and so some failures were simply accepted as inevitable.
(c) Incorrect interpretation of the symptoms (internal damp patches during winter, mould growth, saturated construction, etc.) may have occurred and many roofing contractors were called in to repair 'leaky roofs' following the incorrect assumption that the external waterproofing had failed.
(d) Information dissemination, initial training and in-service training should have created an early end to the problems.

In the situation of energy conservative design, the influence of the architecture on energy consumption is often of equal importance to that of the services. Hence, detailed consideration of energy control during the architectural design is required to the same degree as would be expected during the services design. This spreads the need for respect of energy conservation measures across several responsible persons in the building process:

(a) The client or his representatives who produce the brief and therefore provide the initial direction to the design team.
(b) The architect for his input to the architectural features affecting energy consumption and for the integration of the services and structural design.
(c) The quantity surveyor, cost consultant or building economist for his input on cost information on all the architectural features affecting energy consumption.
(d) The services engineer for his input into the detailed consideration of all the features of the services discussed previously.
(e) The services engineer's cost consultant or quantity surveyor.
(f) The energy consultant for his guidance and advice to all members of the design team.

It is clear that the more demanding projects can only progress successfully if all these separate professional functions are co-ordinated into one team activity directed towards a well-defined common purpose. In several countries, changes have taken place to the traditional method of project management, particularly for the larger projects, initially by introducing the engineer at earlier stages in the project and more recently by having the whole design team managed as a separate professional management responsibility. These project management consultants are often experienced quantity surveyors or architects who act as co-ordinators of the other professionals in the design team. As management is the important function, some project management consultants are management experts who have entered the building industry and have developed their knowledge of the building design process.

Similar developments in management have been taking place on building sites where the main contractor must programme and co-ordinate all the activities of the other contractors within the contract. Some of these contractors have been taking responsibility for the design of the building as well as the construction and so it was inevitable that, with the need for improved building design, they should also employ the building design professionals.

These differing approaches to building design must be contrasted with the opposite extreme. For the simplest of projects, contractors are sometimes selected from a competitive tendering procedure based on a brief which covers only the most basic requirements; some of these requirements may be governed by legislation anyway. Each contractor would be expected to design that section of the work for which he is responsible although occasionally the responsibility for such design work is accepted by equipment manufacturers or systems suppliers, thereby segregating the design work from the hub of the project even further.

Such projects operate quite satisfactorily when there are only a few distinct areas of responsibility such as the erection of a barn for a farm. One responsibility might be accepted for the foundations design, design approval and construction; another for design, erection permission and erection of the framework and fabric of the building; another for the rainwater and surface water drainage. The amount of co-ordination between each discipline is minimal and often only one way; for instance, the designer of the framework and fabric provides dimensions and loads to the designer of the foundations.

The fragmented nature of this design process does not lead to problems in simple projects but as more disciplines are introduced, particularly those which require co-ordination and information exchange with many others, so the need for integration of the separate designers into one team becomes more important.

7.5.2 Design Team Management

The principal design objectives discussed in the sections on ventilation and temperature regulation entail frequent exchange of information between the design team members. Hence, one of the objectives of the management of the design process is to integrate the design team and to provide the conditions which will enable frequent and friendly exchange of information in front of all members so that all views can be expressed.

The management of the design team is an important task, demanding all the skills of management which would be expected from company directors. The creation of a suitable working atmosphere between the members of the team demands the cultivation of harmony between professionals with various depths of understanding which is intrinsic to their training and experience. The team manager does not need to have a detailed understanding of any aspect of the design; any expertise is brought into the design team as required. But he must have an understanding of the requirements of each member and the manner in which they operate so that he can successfully perform the management task of co-ordination and planning. The acceptance of this responsibility relieves the remainder of the team and allows them to concentrate on their particular field of activity.

Each field of activity has evolved from traditional design methods and this has led to some problems over agreements. For example, there does not appear to be any recognised form of agreement over the method of remuneration for an engineer who provides to a design team advice on how energy can be saved. In fact, traditional fee scales based on a percentage of the services costs provide a positive disincentive to a building services engineer to advise on how the loads could be reduced—the harder the engineer works at reducing the loads, so the lower his fees become! The organisation of remuneration should try to prevent these disincentives (see Section 7.1.2).

Evolution from a traditional design approach has also left an inherent problem in the methods of working which do not always ensure that the right information is available at the right time. The manager of the design team must ensure that each appointment of a professional to the team must be accompanied by an agreement over the information required from that professional at each stage in the project.

7.5.3 Initial Stages of the Integrated Design Team Approach

The following approach has some useful merits, irrespective of whether the design team is managed separately or is integrated with the project construction team.

The first design team meeting involves all the principal members. The brief is presented and the principles deriving from the brief are discussed tentatively. Any ideas or experience are brought into the discussions where relevant so that the brief is beginning to create a communication between team members. The meeting then disbands to commence the search for relevant data.

At the next design team meeting the architect presents some sketches which form the vehicle for the discussions between all members of the design team. These initial sketches will have evolved from the architect's interpretation of the brief, the initial discussions and frequent short conversations between the architect and other members of the team since the first meeting. The objective of the meeting is to try to mould the sketch proposals so that all members of the design team feel that it answers the sections of the brief which relate to their expertise. It might be that parts of the proposal do not suitably fit the brief and so these will be resketched, often during the meeting. Occasionally, large portions will be unsuitable and so it is imperative that these initial sketches communicate ideas with the minimum input of time and effort. Throughout this process there may be a need for clarification of some aspects of the brief or more information about the requirements stated in the brief and so occasional communication with the client is necessary.

Design team meetings continue for a pre-programmed period, at the end of which the design becomes 'firm' and the opportunity for larger changes is lost. There is still flexibility within the framework of the design but only so long as any change does not significantly influence other disciplines. Consequently, it is clear that all interdisciplinary influences must be communicated during the initial period of the design. This is the period when the energy consultant must be present to influence the conservation of energy in the design of the building and its services. The services engineer will have been proposing the types of services which are most appropriate for the building options; the eventual decision will be influenced by the economic analyses, the confidence of the design team and the user's ability to understand, accept and operate the systems.

The amount of time allocated to this stage of design must reflect the quantity of material which must be obtained for it to be completed satisfactorily. Experienced energy consultants will be able to direct simple projects or common designs without considerable investment of time for data acquisition. For example, the basic requirements and design implications for an air conditioned office block will often come from experience, although some investment of time will be required for computer simulation studies of the energy consumption for some of the options.

As will be seen later, some decisions during this stage of the project derive from the detailed design requirements which are considered much later in the project. Therefore, the designers who have already completed a similar type of project will be able to progress through the initial stages much more rapidly, with less fruitless exploration of irrelevant material, and guiding other design team members positively and confidently.

Conversely, if it is intended that an office block is to be heated using heat recovered from a computer installation, then more time will be required for the investigation of the methods which might be applicable. Hence, when programming the project, the time allocation to these operations must reflect the degree of complexity of the project and the degree of innovation which may come from new applications of existing techniques. Even though the project may use traditional techniques, there are often aspects over which considerable care must be exercised, such as the trajectory of air jets for air conditioning a large international exhibition centre, the prediction for which cannot be extrapolated confidently from existing information on air jet performance and room air movement (see Section 4.2.4). If the technology is also innovative, additional design time must be allocated to allow for the applied research which will be necessary to ensure successful operation.

As it is usual for the design time to be constrained, the amount of innovation which can be successfully incorporated into each ordinary project is limited. So the process of improving the design of buildings and services for better energy control must take the time needed for the process of gradual innovation on successive projects to reach completion so that a measure of the degree of success can be obtained from its operation.

7.5.4 Later Stages of the Integrated Design Team Approach

After the design has become firm, the energy consultant must influence the details of the design. This incurs the application of all the fundamental principles covered in the early chapters of this book, both to the architecture and the services. This influence may be generalised in the form of guidance and information for the architect when producing more detailed designs, or to such detail that it will be necessary for the energy consultant to produce annotated and dimensioned drawings to indicate the precise nature of his proposals. Computer simulation may be required in either instance.

Regarding the services, the energy consultant will have already influenced the general nature of the services design during the initial stages. More input will be required for the detailed consideration of the performance of the services. For example, during the initial stages of an indoor swimming pool project, some of the sketch stage decisions might have arrived at the generalised conclusions presented in Table 7.1, with the relevant information passed to other design team members. The implications of these decisions often derive from details decided much later in the project. Table 7.2 summarises the decision making process for the energy conserving design of the air handling systems for the indoor swimming pool.

From Tables 7.1 and 7.2 it is clear that the late stage details must be fed back to early stage decisions so that the course for the design to follow smoothly is already provided. Hence, as implied earlier, experience expedites design. When these details are not known, then a search must be made for similar applications so that previous experience can be put to use. When there are no similar applications, it may be necessary to initiate research at an early stage so that the detailed information is available when required. Sometimes, early design decisions must be made in the absence of the detailed data, because time must be provided for the research, and so there may be some inconvenience during the design when changes become necessary. On some occasions, it may be that the design cannot accommodate the results of the research, to the detriment of the project. Consequently, where the results of research are likely to have a significant effect, conclusions must be obtained from the research before design

Table 7.1 Summarised general conclusions and implications of early design decisions for an indoor swimming pool example

Energy consultant's influence on services	Architectural implications	Structural implications	Acoustical implications	Remarks
(a) Gas fuelled engine driven refrigeration compressor to dehumidify pool hall air	Location, volume, access, noise, vibration, weight	Weight, noise, vibration	Noise, vibration	Fundamental decision
(b) Heat recovered from refrigeration compressor for pool water heating	Location, volume	Weight		Minimise length of refrigerant pipework
(c) Heat recovered from refrigeration compressor for pool hall air handling system	Location, volume, route, access, noise, weight	Weight, structural provision for duct routes	Noise	Route and location of air distribution system comes from detailed decisions—information required at this stage obtained from experience or original research
(d) High grade heat recovered from engine for domestic water heating	Location, access, volume, weight	Weight		Domestic hot water storage required. Minimise distribution from refrigeration compressor through storage to end use

Table 7.2 Summarised detailed conclusions and implications for the air handling arrangement for an indoor swimming pool example

Energy consultant's consideration of details	Architectural implications	Structural implications	Acoustical implications	Remarks
(a) Spatial air movement patterns, air velocities, supply and extract location, discharge velocities and temperatures	Diffusers and extract grille locations, appearance	Holes through structure. Structural projections	Noise	The control of the air quality in the pool hall is dependent on the details in this part of the design—the general implications must be communicated to other team members during initial project stages
(b) Air duct sizes, velocities, resistance to air flow	Route, size	Holes through structure or structural provisions for duct routes	Noise breakout, ductborne noise from central plant	

(c) Air handling unit size and air flow resistance caused by filters, coils, dampers, ducts, grilles and diffusers. Energy consumption by main air handling fans	Size, location, noise, vibration	Weight, noise, vibration	Noise, vibration, ductborne transmission	
(d) Selection of refrigerant evaporating and condensing temperatures for the air handling unit coils, coil sizes, air flow resistance				Fundamental information necessary for obtaining the coefficient of performance of the refrigeration compressor—required during the initial stages for the economic analyses

proceeds beyond the stage when the conclusions cannot be incorporated easily.

7.5.5 Decisions Regarding the Thermal Performance of the Building

During early stages of design, the energy consultant must give the design team some indication of the most economic thermal transmittance value for the building envelope. This is not necessarily the performance which will be required eventually, but is intended to provide guidance so that the design team do not waste time with building fabric which is inappropriate.

The economics of the thermal transmittance values cannot be assessed during early stages as the details are not decided. Hence, these early decisions must be made from experience and intuition, with some foresight as to what is likely to be proved to be economically viable in later stages. Using the example of project decisions in Tables 7.1 and 7.2, the economics of the thermal transmittance of the building envelope cannot be assessed until the unit cost of recovered energy is obtained and this depends on the selection of a fuel and system, and the performance of the system.

Another fundamental building fabric decision during early stages is whether the building shall be designed for a slow thermal response or for a fast response. This coarse approach is usually adequate at initial stages of the project as it provides the design team with guidance about the internal partitions, floors, ceilings and the location of insulation in the envelope. At early stages, some guidance regarding the amount of glazing will be useful. The brief may stipulate the daylighting requirements, in which case the glazing design will be constrained, otherwise the influence of glazing on each of the building facades must be considered for heat loss, heat gain, appearance, daylighting, sunlighting, views and the psychological relationships between occupants and the external environment. Initial general guidance will provide direction for detailed consideration of the glazing on each facade.

At later stages in the project, it will be necessary to consider in more detail the materials and glazing which might be chosen and their influence on the overall thermal transmittance and the thermal response. It is at these more detailed stages that the computer simulations can provide assistance in choosing between one option and another. For example, the selection of surface materials with different thermal penetration coefficients can be analysed for their effect on energy consumption in intermittently heated buildings. The selection of the economic thickness of insulation and where it is located can also be analysed (see Section 6.4.2). The effects of such options on energy consumption can be compared to the changes in capital cost and construction time caused by selecting an option. These analyses require information from many members of the design team so that the economics can be assessed.

7.5.6 *Mechanical Services Decisions*

General guidance from the energy consultant during early stages must continue to the consideration of details as the project progresses and the services design approaches the final stages. Throughout the design it will be necessary to apply the approach given in previous chapters so that all aspects of the design reflect an energy conservative approach. This procedure applies whatever the complexity of the services.

Consideration of the details of the mechanical services design does not usually have implications on the other members of the design team. The process is one of considering the design options and then making a decision based on economic analyses or other influences, but usually affecting only the services and energy consumption. For example, the selection of discharge air temperatures from an air handling system will influence the air temperature gradients in the space and therefore affect the energy lost or gained through the fabric. As the difference in air temperature between the space and the discharge from the air handling unit decreases, so the plant size increases and the energy consumed by the air handling fans increases. Hence, this is an optimisation process; the reduced energy transfer through improved air temperature control in the space must be justified by the increased capital and running costs of the services. This process becomes very complex when the performance of refrigeration systems enters the optimisation.

The results of these design options are generally confined to the mechanical services and have little bearing on the other professions. One exception is the detailed dimensions and appearance of all items affecting the requirements of the designer of the building interior, as this information will be required for co-ordination.

7.6 SMALL PROJECTS

As there is often no design team of the type discussed previously for small projects, it is important to ensure that any expertise is brought into the design process at the relevant stage. For example, some fume cupboards were required for an existing laboratory. The laboratory staff obtained competitive quotations from three companies, one of which was successful and completed the installation. Afterwards, it was discovered that there was no consideration of the economics of energy conservation and moreover there was no provision for fresh air to gain access to the laboratory to replace the air exhausted through the fume cupboards. Consequently, the systems were unable to achieve the intended performance. (Incidentally, when the laboratory staff were informed that the fume cupboard operating costs, due to the discharge of heated air, were equivalent to the heat energy consumption of an average house, they decided to operate the new supply air plant and fume cupboards only when necessary, instead of continuously as they had done previously).

In small projects there is usually no need for the full expertise of all the professionals in the building design team, but the engagement of a building services engineer or energy consultant may prevent such problems arising. This applies specifically to industrial services where the quantities of energy consumed are relatively large, but the same principles can be applied to the design of a single dwelling which may also benefit from the introduction of expertise on energy conservation during the design process. In addition, the appointment of one design team professional on a project will often help to identify other problem areas which otherwise may have remained unnoticed.

The appointment of the appropriate consultants should assist with the preparation and development of the brief, the preparation of a design and specification of materials, components or systems, the selection of suitable contractors for competitive tendering and the analysis of the tenders and prices. It may also prove beneficial to retain the consultants for inspection of the work to check that everything is satisfactorily installed and commissioned.

A consultant can sometimes be selected from the lists often kept by professional institutions and organisations (e.g. [7.4, 7.5, 7.6]). Contact with the appropriate institution will often lead to the names of several consultants who might be able to assist with a project. An interview with a consultant before appointment is usually beneficial as it provides an opportunity to assess the consultants' field of operation, previous experience and views on the project in question.

This situation presents a problem with small projects for small firms with little turnover. Funds for energy conservation are usually much smaller compared to the large firms, and less time and effort can be expended on energy studies. The appointment of an energy consultant is not so fundamentally important as the appointment of, say, a structural engineer. In fact, usually there is no legislation to make the appointment of an energy consultant mandatory. In projects where this expertise is not brought in, the responsibility for economic design of the thermal performance of the building and its services rests with the remainder of the design team. As a consequence, it is important that the designers have a general understanding of the basic principles of thermal energy conservation so that the information may be incorporated into the design. Unfortunately, it is not uncommon for the pressures on a designer to be so burdensome that there is no time or incentive to place any effort in this direction. However, the designers must recognise the complexity of a new project and, if it is beyond their scope, appropriate expertise should be brought in at the right time.

7.7 CONCLUSIONS

Usually, the subject of energy conservation is not directly of interest to the average consumer, except when there are moral obligations or when the supply of energy is strictly limited. Instead, any incentive to conserve energy normally derives from the financial savings and so the increasing values of energy have naturally encouraged more consideration of energy conserving measures.

Buildings and services represent one area where energy conservation can produce substantial benefits. Yet the blind pursuit of energy conservation and the use of sophisticated technology as a priority during design is unlikely to produce an acceptable solution to a brief because of the complex nature of the design process and the many compromises that are intrinsic to the design of buildings and services.

The traditional method of one profession designing the building to achieve passive regulation of temperatures and then another profession designing the services at a later date to achieve active regulation of temperatures, has had to cease. Instead, it is necessary to consider these thermal processes as one integrated yet complex series of continuing events. This requires an understanding of sufficient depth that the subject of energy conservation can be regarded as a separate professional discipline. As with any other discipline, it is not necessary for each member of a building design team to understand the detailed aspects of energy conservation but this book has attempted to show that an understanding of some of the general principles and the fundamentals behind these generalities is useful. The precise details should be left to an energy consultant so that the other design team members can concentrate on their particular interests, although frequent co-ordination between the energy consultant and all design team members is essential. This need for frequent communication between the energy consultant and the other members of the design team operates more successfully in the integrated design team approach.

It is generally accepted that the value of energy will continue to increase and so there is likely to be a continuing trend towards more energy conservation and changes to other sources of energy as some energy sources become more expensive than others. Consequently, there is also a continuing need for the building design team, in considering the complex function of the building and its services, to use the current technical understanding and design facilities to produce designs which are the most economic balance of the passive and active forms of thermal control.

REFERENCES

7.1 MERRETT, A. J. and SYKES, A.: *The Finance and Analysis of Capital Projects*. Longmans, London, 1963.

7.2 CHARTERED INSTITUTION OF BUILDING SERVICES: *Building Energy Code: Part III, Guidance towards Energy Conserving Operation of Building and Services*. CIBS Publications, London, 1979.

7.3 BUILDING RESEARCH ESTABLISHMENT WORKING PARTY: *Energy Conservation: A Study of Energy Consumption in Buildings and Possible Means of Saving Energy in Housing*. Building Research Establishment Current Paper, CP 56/75, 1975.

7.4 THE INSTITUTE OF ENERGY, 18 Devonshire Street, Portland Place, London.

7.5 CHARTERED INSTITUTION OF BUILDING SERVICES, Delta House, 222 Balham High Road, London SW12 9BS.

7.6 THE AMERICAN SOCIETY OF HEATING, REFRIGERATING AND AIR-CONDITIONING ENGINEERS, 345 East 47th Street, New York.

Index

17.000